AF893291

# Methods in Molecular Biology

*Series Editor*
**John M. Walker**
**School of Life and Medical Sciences,**
**University of Hertfordshire,**
**Hatfield, Hertfordshire, UK**

For further volumes:
http://www.springer.com/series/7651

For over 35 years, biological scientists have come to rely on the research protocols and methodologies in the critically acclaimed *Methods in Molecular Biology* series. The series was the first to introduce the step-by-step protocols approach that has become the standard in all biomedical protocol publishing. Each protocol is provided in readily-reproducible step-by-step fashion, opening with an introductory overview, a list of the materials and reagents needed to complete the experiment, and followed by a detailed procedure that is supported with a helpful notes section offering tips and tricks of the trade as well as troubleshooting advice. These hallmark features were introduced by series editor Dr. John Walker and constitute the key ingredient in each and every volume of the *Methods in Molecular Biology* series. Tested and trusted, comprehensive and reliable, all protocols from the series are indexed in Pub Med.

# Pathogenic *Yersinia*

## Methods and Protocols


Edited by

**Viveka Vadyvaloo**

*Paul G Allen School for Global Animal Health, Washington State University, Pullman, WA, USA*

**Matthew B. Lawrenz**

*Department of Microbiolgy and Immunology and Center for Predictive Medicine in Biodefense and Emerging Infectious Diseases, University of Louisville School of Medicine, Louisville, KY, USA*



Humana Press

*Editors*
Viveka Vadyvaloo
Paul G Allen School for Global Animal Health
Washington State University
Pullman, WA, USA

Matthew B. Lawrenz
Department of Microbiolgy and Immunology and Center for Predictive Medicine in Biodefense and Emerging Infectious Diseases
University of Louisville School of Medicine
Louisville, KY, USA

ISSN 1064-3745 ISSN 1940-6029 (electronic)
Methods in Molecular Biology
ISBN 978-1-4939-9540-0 ISBN 978-1-4939-9541-7 (eBook)
https://doi.org/10.1007/978-1-4939-9541-7



This Humana imprint is published by the registered company Springer Science+Business Media, LLC, part of Springer Nature.
The registered company address is: 233 Spring Street, New York, NY 10013, U.S.A.

# Preface

There are three major species of the genus *Yersinia* that cause human disease. These are *Y. enterocolitica*, *Y. pseudotuberculosis*, and *Y. pestis*. While the former two species are etiological agents of the mild and resolvable intestinal disease referred to as yersiniosis, *Y. pestis* causes the fulminant disease, plague. The intestinal pathogens can survive within environmental niches, such as soil and water, and are often contracted by eating contaminated food. In contrast, *Y. pestis* is more sensitive to environmental conditions and is maintained within a mammalian-flea infectious cycle, with human transmission primarily occurring through contact with infected fleas. However, *Y. pestis* is also able to be transmitted via inhalation of contaminated aerosols, allowing for the possibility of direct human-to-human transmission.

The purpose of this book is to deliver a compendium of detailed protocols to the research community to aid in the investigation and discovery of *Yersinia* virulence mechanisms using important in vivo and in vitro infection models. These models have led to major advances in the field and in understanding *Yersinia* host-pathogen interactions. However, the original scientific publications utilizing *Yersinia* experimental models often do not provide a complete step-by-step protocol, or list of materials, that are critical to achieve success early during the experimental optimization process. Therefore, the protocols within this volume represent the most current methods from expert researchers in the *Yersinia* field and provide comprehensive procedures and advice, such that a novice in the field might be able to effortlessly develop effective and reproducible protocols and assays.

While many different animal models have been employed by *Yersinia* researchers, the mouse has been the most extensively used and is the most accessible to those new to the field. Therefore, we have focused on providing detailed protocols to establish infection in murine models of plague and yersiniosis. Mouse models for plague are intended to mimic the key manifestations of human plague infection. Bubonic plague results from flea transmission, in which the flea deposits the bacteria within the dermis during feeding, and can be mimicked in the laboratory via flea transmission or intradermal injection. Primary pneumonic plague arises from inhalation of *Y. pestis*, which can be modeled by either intranasal and aerosol instillation. Because yersiniosis ensues after acquisition of *Y. enterocolitica* or *Y. pseudotuberculosis* via the fecal-oral route, oral infection of mice is the most appropriate model to study this disease in the laboratory. Importantly, each animal protocol in this volume includes considerations for safe and humane laboratory animal handling practices.

After infection, monitoring bacterial proliferation and the host immune response is paramount in understanding pathogenesis. *Yersinia* researchers have developed several cutting-edge methods for visualizing bacterial dissemination and encounters with immune cells both in vivo and in intact fixed host tissue. Multiple chapters in this volume are specifically devoted to descriptions of animal and tissue preparation for imaging and associated utilization and advantages of intravital microscopy, laser scanning microscopy, and bioluminescent small animal imaging to understanding the infection process. Moreover, both pathogen and host undergo important changes in gene expression during the infection process to coordinate virulence gene and immune responses, respectively. Therefore, approaches to monitor transcription of both the pathogen and host, including the simultaneous host-pathogen transcriptomic analysis known as "dual RNA-seq," are described.

The key to successful host colonization by the *Yersinia* arises from their ability to evade clearance by macrophages and neutrophils. Furthermore, immune evasion hinges largely on the production of a major virulence factor, the Ysc type three secretion system (T3SS), and deployment of Yop effector proteins via this system directly into host cells to inhibit host signaling pathways. As such, detailed protocols describing in vitro infection of neutrophils and macrophages, targeting of immune cells by the T3SS in vivo, and methods to monitor changes in host signaling pathways, including the inflammasome, are discussed in several chapters.

In addition to vertebrate infection models for plague, invertebrate infection models have been developed to understand the insect vector host-*Y. pestis* interaction. Therefore, chapters describing the use of insect vectors in plague research are an important and unique aspect of this volume of Methods in Molecular Biology. Because the flea is the natural insect host for plague, protocols to establish flea infection and monitor bacterial proliferation within the flea are included. However, *Drosophila* has also recently been developed as an additional invertebrate plague model, providing a powerful genetically tractable model to uncover the insect vector response to *Y. pestis* infection, and is included in a separate chapter.

Together, the chapters in this volume describe the significant infection models used to understand the biology of human diseases caused by *Yersinia* pathogens. We would like to thank Dr. John Walker for the opportunity to compile these protocols into a single source. We are also grateful to the contributing authors for generously sharing the protocols that have been painstakingly developed, optimized, and mastered in their own labs.

*Pullman, WA, USA* *Viveka Vadyvaloo*
*Louisville, KY, USA* *Matthew B. Lawrenz*

# Contents

## Part III Invertebrate Models of Infection (In Vivo and In Vitro)



## Part IV *Yersinia* Interaction with Immune Cells and Immune Signaling

## Contributors


DEBORAH M. ANDERSON • *Department of Veterinary Pathobiology, University of Missouri, Columbia, MO, USA; The Laboratory for Infectious Disease Research, University of Missouri, Columbia, MO, USA*

PAUL E. ANDERSON • *Department of Veterinary Pathobiology, University of Missouri, Columbia, MO, USA; The Laboratory for Infectious Disease Research, University of Missouri, Columbia, MO, USA*

VICTORIA AUERBUCH • *Department of Microbiology and Environmental Toxicology, University of California, Santa Cruz, Santa Cruz, CA, USA*

MICHAEL BECKSTETTE • *Department of Molecular Infection Biology, Helmholtz Centre for Infection Research, Braunschweig, Germany*

JAMES B. BLISKA • *Department of Molecular Genetics and Microbiology, Center for Infectious Diseases, Stony Brook University, Stony Brook, NY, USA; Department of Microbiology and Immunology, Geisel School of Medicine at Dartmouth, Hanover, NH, USA*

IGOR E. BRODSKY • *Department of Pathobiology, University of Pennsylvania School of Veterinary Medicine, Philadelphia, PA, USA*

ELISABETH CARNIEL • *Yersinia Research Unit/Yersinia National Reference Laboratory/WHO Collaborating Center, Institut Pasteur, Paris, France; Centre Pasteur du Cameroun, Yaoundé, Cameroon*

TIAGO R. D. COSTA • *Department of Molecular Biology, Umeå University, Umeå, Sweden; Umeå Centre for Microbial Research, Umeå University, Umeå, Sweden; Department of Life Sciences, MRC Centre for Molecular Bacteriology and Infection, Imperial College, London, UK*

ANNE DERBISE • *Yersinia Research Unit/Yersinia National Reference Laboratory/WHO Collaborating Center, Institut Pasteur, Paris, France*

PETRA DERSCH • *Department of Molecular Infection Biology, Helmholtz Centre for Infection Research, Braunschweig, Germany; Institute of Infectiology, University of Münster, Münster, Germany*

OLIVIER DUSSURGET • *Yersinia Research Unit/Yersinia National Reference Laboratory/WHO Collaborating Center, Institut Pasteur, Paris, France; Université de Paris, Sorbonne Paris Cité, Paris, France*

TOMAS EDGREN • *Department of Molecular Biology, Umeå University, Umeå, Sweden; Umeå Centre for Microbial Research, Umeå University, Umeå, Sweden; Department of Medical Biochemistry and Microbiology, Uppsala Biomedical Center, Uppsala University, Uppsala, Sweden*

KARA R. EICHELBERGER • *Department of Microbiology and Immunology, University of North Carolina at Chapel Hill, Chapel Hill, NC, USA*

SALAH I. FARAG • *Department of Molecular Biology, Umeå University, Umeå, Sweden; Umeå Centre for Microbial Research, Umeå University, Umeå, Sweden*

ALYSSA C. FASCIANO • *Immunology Graduate Program, Sackler School of Biomedical Science, Boston, MA, USA; Department of Molecular Biology and Microbiology, Tufts University School of Medicine, Boston, MA, USA*

MATTHEW S. FRANCIS • *Department of Molecular Biology, Umeå University, Umeå, Sweden; Umeå Centre for Microbial Research, Umeå University, Umeå, Sweden*

MONIKA K. FRANCIS • *Department of Molecular Biology, Umeå University, Umeå, Sweden; Umeå Centre for Microbial Research, Umeå University, Umeå, Sweden*
WILLIAM E. GOLDMAN • *Department of Microbiology and Immunology, University of North Carolina at Chapel Hill, Chapel Hill, NC, USA*
RODRIGO J. GONZALEZ • *Department of Immunology, Harvard Medical School, Boston, MA, USA*
ANN KATHRIN HEROVEN • *Department of Molecular Infection Biology, Helmholtz Centre for Infection Research, Braunschweig, Germany*
B. JOSEPH HINNEBUSCH • *Laboratory of Bacteriology, Rocky Mountain Laboratories, NIAID, NIH, Hamilton, MT, USA*
DIANA HOOKER-ROMERO • *Department of Microbiology and Environmental Toxicology, University of California, Santa Cruz, Santa Cruz, CA, USA*
MARIA KUSMIEREK • *Department of Molecular Infection Biology, Helmholtz Centre for Infection Research, Braunschweig, Germany*
MATTHEW B. LAWRENZ • *Department of Microbiology and Immunology, Center for Predictive Medicine for Biodefense and Emerging Infectious Diseases, University of Louisville School of Medicine, Louisville, KY, USA*
ATHENA LEMON • *Paul G. Allen School for Global Animal Health, Washington State University, Pullman, WA, USA*
YINON LEVY • *Department of Biochemistry and Molecular Genetics, Israel Institute for Biological Research, Ness-Ziona, Israel*
ANNA B. LUDLOW • *Department of Veterinary Pathobiology, University of Missouri, Columbia, MO, USA; The Laboratory for Infectious Disease Research, University of Missouri, Columbia, MO, USA*
EMANUELLE MAMROUD • *Department of Biochemistry and Molecular Genetics, Israel Institute for Biological Research, Ness-Ziona, Israel*
MELANIE M. MARKETON • *Department of Molecular Biology and Microbiology, Tufts University School of Medicine, Boston, MA, USA*
ANNE L. MCCABE • *Department of Molecular Biology and Microbiology, Tufts University School of Medicine, Boston, MA, USA*
JOAN MECSAS • *Department of Molecular Biology and Microbiology, Tufts University School of Medicine, Boston, MA, USA*
NATASHA P. MEDICI • *Department of Molecular Genetics and Microbiology, Center for Infectious Diseases, Stony Brook University, Stony Brook, NY, USA; Department of Microbiology and Immunology, Geisel School of Medicine at Dartmouth, Hanover, NH, USA*
GIANG T. NGUYEN • *Immunology Graduate Program, Sackler School of Biomedical Science, Boston, MA, USA; Department of Molecular Biology and Microbiology, Tufts University School of Medicine, Boston, MA, USA*
AARON M. NUSS • *Department of Molecular Infection Biology, Helmholtz Centre for Infection Research, Braunschweig, Germany*
RACHEL M. OLSON • *Department of Veterinary Pathobiology, University of Missouri, Columbia, MO, USA; The Laboratory for Infectious Disease Research, University of Missouri, Columbia, MO, USA*
CASSANDRA D. PAULING • *Department of Veterinary Pathobiology, University of Missouri, Columbia, MO, USA; The Laboratory for Infectious Disease Research, University of Missouri, Columbia, MO, USA*

Roger D. Pechous • *Department of Microbiology and Immunology, University of Arkansas for Medical Sciences, Little Rock, AR, USA*
Javier Pizarro-Cerdá • *Yersinia Research Unit/Yersinia National Reference Laboratory/ WHO Collaborating Center, Institut Pasteur, Paris, France*
Amanda R. Pulsifer • *Department of Microbiology and Immunology, Center for Predictive Medicine for Biodefense and Emerging Infectious Diseases, University of Louisville School of Medicine, Louisville, KY, USA*
Janelle Sagawa • *Paul G. Allen School for Global Animal Health, Washington State University, Pullman, WA, USA*
Leah Schwiesow • *Department of Molecular, Cell and Developmental Biology, University of California, Santa Cruz, Santa Cruz, CA, USA*
Jeffrey G. Shannon • *Laboratory of Bacteriology, Rocky Mountain Laboratories, NIAID, NIH, Hamilton, MT, USA*
Amelia Silva-Rohwer • *Paul G. Allen School for Global Animal Health, Washington State University, Pullman, WA, USA*
Viveka Vadyvaloo • *Paul G. Allen School for Global Animal Health, Washington State University, Pullman, WA, USA*
Yaron Vagima • *Department of Biochemistry and Molecular Genetics, Israel Institute for Biological Research, Ness-Ziona, Israel*
Tiva T. VanCleave • *Department of Microbiology and Immunology, Center for Predictive Medicine for Biodefense and Emerging Infectious Diseases, University of Louisville School of Medicine, Louisville, KY, USA*
Yahan Wei • *Department of Microbiology and Environmental Toxicology, University of California, Santa Cruz, Santa Cruz, CA, USA*
Joshua L. Willix • *Department of Veterinary Pathobiology, University of Missouri, Columbia, MO, USA; The Laboratory for Infectious Disease Research, University of Missouri, Columbia, MO, USA*
Erin E. Zwack • *Department of Pathobiology, University of Pennsylvania School of Veterinary Medicine, Philadelphia, PA, USA; Department of Microbiology, New York University School of Medicine, New York, NY, USA*

# Part I

## Mouse Models for *Yersinia* Infection

# Chapter 1

# An Intradermal Model for *Yersinia pestis* Inoculation

**Rodrigo J. Gonzalez**


## Abstract

The dermis and the subcutaneous space vary in many fundamental characteristics, which include composition of lymphatic vessels, density of blood vasculature, and cells of the immune response. Traditional approaches employ the subcutaneous space as the preferred layer of the skin to inoculate *Yersinia pestis* for bubonic plague studies. Because fleas transmit *Y. pestis* in nature, and because these insects target the dermal layer of the skin, an intradermal model of infection is more biologically relevant than a subcutaneous model. Among many features, the use of an intradermal model results in robust and reproducible colonization of lymph nodes, blood, and deeper tissues. Remarkably, intradermal inoculation in the murine ear pinna also allows for the study of cutaneous infection without severely disrupting the architecture and physiology of the skin.


**Key words** Intradermal inoculation, Dermis, Ear pinna, Skin, Bubonic plague, *Yersinia pestis*

## 1 Introduction

*Yersinia pestis* can survive in the digestive tract of a flea and can be transmitted to a host when the insect probes to take a blood meal [1, 2]. Consequently, and while the physiology of transmission through a fleabite is not fully understood, the flea's mouth parts can be thought of as inoculation needles that deliver *Y. pestis* into the skin of a host.

In the laboratory, the use of fleas to inoculate mice with *Y. pestis* is challenging. Inoculation by fleas makes it impossible to control the inoculum to be delivered into a mouse [3]. Therefore, using fleas to inoculate mice results in high variability of infection between individuals. Moreover, fleas are not easy to manipulate, especially in a biosafety level three (BSL3) facility. For these reasons, the use of hypodermic needles to inoculate mice is inevitable for the study of bubonic plague.

Even when hypodermic needles can replace the use of fleas, it is very important to think about the implications of *Y. pestis* being transmitted through a fleabite in nature. Traditionally, inoculations

Viveka Vadyvaloo and Matthew B. Lawrenz (eds.), *Pathogenic Yersinia: Methods and Protocols*, Methods in Molecular Biology, vol. 2010, https://doi.org/10.1007/978-1-4939-9541-7_1, 

of *Y. pestis* in mouse models of bubonic plague are done through subcutaneous injections [4–7]. However, fleas have mouthparts that penetrate into the dermis [2, 8, 9], and not into the much deeper subcutaneous space. This notion is fundamental because the different layers of the skin vary considerably in elements such as blood vasculature, populations of cells of the immune response, and, importantly for bubonic plague, lymphatic vasculature [10–12]. The dermis is particularly rich in afferent lymphatic vasculature. These afferent lymphatic vessels move lymph toward draining lymph nodes. More importantly, this vasculature is mostly composed of terminal lymphatics, blind-ended lymphatic vessels that directly collect lymph from a tissue [13]. On the other hand, the subcutaneous space contains thicker and less permeable lymphatic vessels, known as collector lymphatic vessels, or simply collectors [13, 14]. Collectors are thought to be less efficient in the uptake of particles associated with lymph than terminal lymphatics. *Y. pestis* most likely evolved in the context of the dermis, not the subcutaneous space. Accordingly, bacteria–host interactions that are physiologically relevant should be expected to occur when infection is studied in the context of an intradermal and not a subcutaneous model of infection. Therefore, it is not surprising that key parameters of pathogenesis, such as bacterial dissemination kinetics or host survival, vary considerably between models using the dermis or the subcutaneous space as the site of inoculation [10].

Intradermal inoculation can be achieved virtually in any anatomical part of the mouse. However, the most appropriate site for the procedure is the ear pinna. This is because the ear pinna is unique in that the dermis lies directly on the cartilage that divides the ventral and dorsal leaflets that form the auricular structure. Consequently, there is virtually no subcutaneous space in the ear pinna. While inoculation in the dermis at other anatomical sites could hypothetically be done, it is practically impossible to demonstrate the needle did not penetrate beyond the dermis into the subcutaneous space. Therefore, other anatomical sites should be avoided when considering intradermal models of infection.

Intradermal inoculations have been used extensively in virology and parasitology. It was from a modified protocol for inoculation of *Leishmania* that an intradermal inoculation model was developed to be used in modern studies of bubonic plague [7]. The method presented in this chapter varies from other methods mainly in that the inoculum is contained in a very small volume (~2 μL) and that it comprises relatively low bacterial numbers. The use of lower volumes makes the method more technically challenging, but it reduces the possibility of confounding effects due to large volumes disrupting the tissue [10]. The method is divided into three parts: preparation of the inoculum, inoculation into the ear pinna, and assessment of bacterial burden in the skin and other tissues. Each section is described assuming the procedures will be conducted following aseptic technique and inside a biosafety cabinet in a BSL3 facility. Other than brief comments on specific procedures that might be of considerable risk, no reference to common practices in a BSL3 facility will be made.

## 2 Materials

### 2.1 Preparation of the Inoculum

1. Brain–heart infusion (BHI) agar plates.
2. Inoculation loops.
3. BHI broth.
4. Microbial culture rotator (i.e., microbial culture roller drum) to provide aeration to cultures.
5. Vortex mixer.
6. $OD_{600}$ spectrophotometer and cuvettes.
7. Phosphate-buffered saline (PBS).
8. 15 mL conical tubes for serial dilutions
9. Incubator at 26 °C.

### 2.2 Inoculation into the Ear Pinna

1. Automatic syringe pump with pedal.
2. Glass syringe.
3. Inoculation stage with autoclaved absorbent paper.
4. Anesthetics.
5. Double-sided adhesive tape.
6. High strength adhesive tape (autoclave adhesive tape recommended).
7. Tweezers.
8. Autoclaved paper towels (to clean the tip of the needle).
9. Butterfly needle (at least 27G).

### 2.3 Assessment of Bacterial Burden in the Skin and Other Tissues

1. Dissecting tools.
2. Bead homogenizer machine.
3. Bead homogenizer vials.
4. Bead homogenizer beads (zirconia beads recommended).
5. Brain–heart infusion agar plates.
6. Ice.
7. Homogenizing pouches.
8. Homogenizing rod.
9. Microcentrifuge tubes with PBS for serial dilutions.
10. Incubator at 26 °C.
11. PBS.

## 3 Methods

It is assumed the researcher is familiar with aseptic technique and general BSL3 procedures. Consequently, these procedures will not be described in the following methods.

### 3.1 Preparation of the Inoculum

1. Streak a BHI agar plate with *Y. pestis* from frozen glycerol culture stocks kept at −80 °C.
2. Incubate the plate at 26 °C until bacterial growth is evident (approximately 48 h). The plate can be stored at 4° for up to 1 week.
3. Prepare a glass test tube with 2 mL BHI broth. With an inoculation loop, scrape off bacteria from the area of the plate with most confluent growth (*see* **Note 1**) and transfer into the test tube.
4. Incubate the inoculated test tube at 26 °C (to mimic the temperature at which *Y. pestis* grows in the flea) for approximately 15 h providing aeration by placing it in a microbial culture roller drum (*see* **Notes 2** and **3**).
5. After incubation, unless the rationale of the experiment suggests otherwise, mix vigorously with a vortex mixer to incorporate the biofilm ring growing on the walls of the tube into the culture (*see* **Note 3**).
6. Measure the optical density at 600 nm ($OD_{600}$) of the culture with a spectrophotometer.
7. Make a dilution of the original culture (in PBS) to obtain a culture at a concentration of 0.5 $OD_{600}$/mL in a total volume of 10 mL. This will be considered dilution $10^{-1}$ and will contain 5 $OD_{600}$, or approximately $1.5 \times 10^9$ colony forming units (CFU) as 1 $OD_{600} = 3 \times 10^8$ CFU.
8. Take a 15 mL conical tube with 9 mL PBS and add 1 mL of dilution $10^{-1}$ to obtain dilution $10^{-2}$ (*see* **Notes 4** and **5**).
9. Continue making 1:10 dilutions until dilution $10^{-4}$ is achieved.
10. Take 2 μL of the $10^{-4}$ dilution, predicted to have approximately 150 CFU/μL, and mix this volume with 198 μL PBS in a microcentrifuge tube. Plate 100 μL of the mix on BHI agar and incubate for 48 h at 26 °C. Do this procedure in triplicate. Count the number of CFU on each plate and multiply by two to quantify the inoculum (CFU in 2 μL).

### 3.2 Inoculation Into the Ear Pinna

1. Prepare an injection stage by securing a 50 mL conical tube onto an empty tip box with adhesive tape (*see* Fig. 1). Make sure the conical tube is tightly secured so that it does not move in any direction.

**Fig. 1** Stage to facilitate intradermal inoculation into the mouse ear pinna. The stage is constructed with a conical tube attached with adhesive tape to an empty tip box. Each panel shows different angles of the stage to appreciate how it is constructed

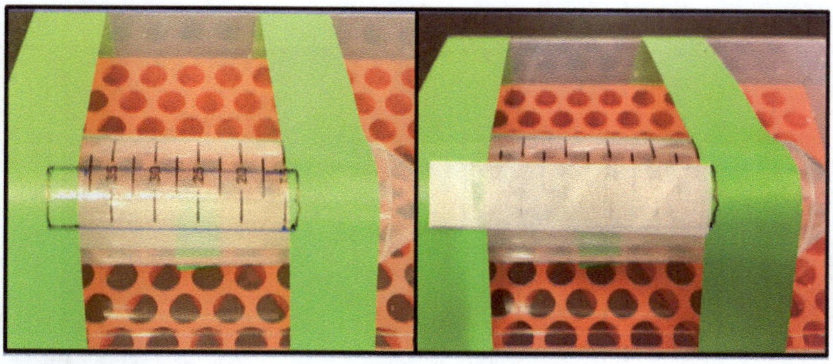

**Fig. 2** Transparent adhesive double-sided tape (delineated with blue marker for visualization purposes) adhered to the conical tube (left). The use of double-sided tape facilitates the adhesion of adhesive autoclave tape (placed with the sticky side up) to the conical tube (right). The ear pinna will be secured to the conical tube through the sticky side of the adhesive autoclave tape

2. Place a piece of adhesive double-sided tape on the conical tube and place a piece of adhesive autoclave tape on top (adhesive side up, *see* Fig. 2 and **Note 6**).
3. From the tubes that contain serial bacterial dilutions made in **step 9** of the previous section, take the tube predicted to contain bacteria at a concentration of approximately 150 CFU/μL (*see* **Notes 7** and **8**), mix vigorously (*see* **Note 5**) and empty into a small petri dish or short tube to facilitate loading of the glass syringe without contaminating the external walls of the syringe. Load the syringe and secure the Luer plastic adapter of the butterfly needle by twisting onto the Luer-lock fitting of the glass syringe (*see* **Note 9**).
4. Eliminate air bubbles from the syringe and tubing of the butterfly needle and secure the syringe into the automatic syringe pump device (*see* **Note 10**).

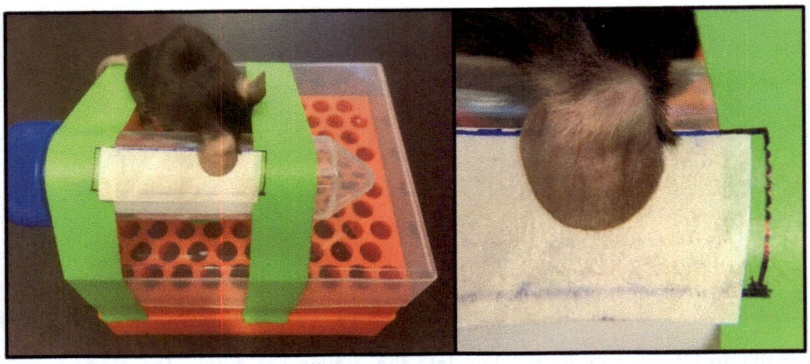

**Fig. 3** An anesthetized mouse positioned in such a way that the ear pinna can easily be placed on top of the conical tube of the stage (left). Approximately two-thirds of the ear are placed on the adhesive autoclave tape (right)

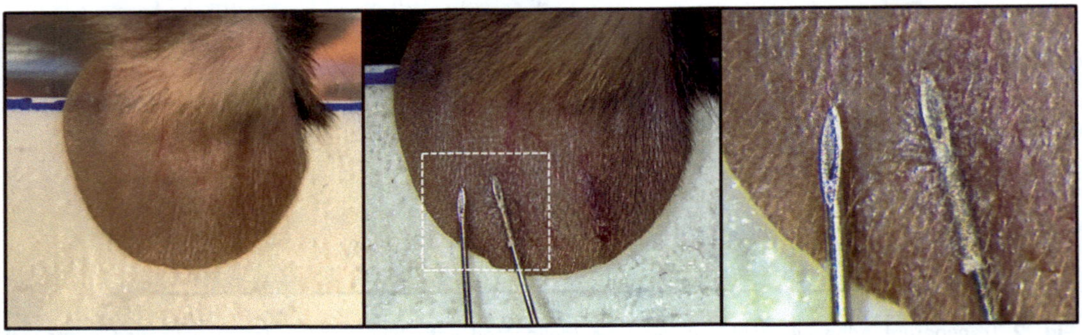

**Fig. 4** Ear pinna adhered to autoclave tape (left panel). Notice the pinna is completely flat, with no wrinkles. After the pinna was in place, a needle was inserted into the dermis (right needle) and another needle was placed next to it (not inserted, left needle) as a reference to appreciate how light reflection against the metallic needle is diminished by the skin (center and right panel). The dotted line delineates the area of the image that is enlarged in the right panel

5. Take a fully anesthetized mouse (a ketamine–xylazine cocktail is recommended to keep the mouse completely immobile) and press the ventral side of the ear pinna firmly against the autoclave tape on the conical tube (*see* Fig. 3) in such a way that the ear is completely extended as shown in Fig. 4 (*see* **Note 11**).
6. Hold the butterfly needle and uncap it with the help of tweezers (*see* **Note 12**).
7. Press the footswitch (injection pedal) as many times as it is necessary to eliminate air bubbles from the tubing. Press the foot switch a few more times to ensure that the system is generating a drop (inoculum) of consistent size at the tip of the needle each time. Use a piece of autoclaved absorbent paper to clean each new drop (*see* **Note 13**).
8. With the opposite hand, move the injection stage with the mouse toward the hand holding the needle. Keep the hand

holding the needle as still as possible as movement will result in deformation of the tubing and, consequently, changes in internal pressure will affect the volume to be injected.

9. Insert the needle into the skin at the edge of the ear, bevel up, slowly and almost completely parallel to the skin, as shown in Fig. 4. Push the needle in so the tip of the needle moves away from the edge of the ear and then retract slightly so there is a space in the tissue for the inoculum to be injected (*see* **Note 14**).
10. Press the footswitch to inject the inoculum and retract the needle completely. A small and perfectly round wheal should be clearly visible for the next minute or so. No liquid should come out of the injection site, as the entire inoculum should be contained in the wheal (*see* **Notes 15** and **16**).
11. Once all mice are inoculated, collect a sample of the inoculum by activating the footswitch and depositing the 2 μL droplet at the tip of the needle in a microcentrifuge tube containing 198 μL PBS. Mix thoroughly with a vortex mixer and plate 100 μL on a BHI plate. Incubate at 26 °C for 48 h, count the number of CFU growing on the plate and multiply it by two. This will be the inoculum, that is, the number of CFU contained in a volume of 2 μL. Do this procedure in triplicate and report the inoculum as the average CFU of the three plates (*see* **Note 17**).

### 3.3 Assessment of Bacterial Burden in Different Tissues

This section describes procedures for assessing total bacterial burden in the ear, lymph nodes, spleen, and blood. These procedures do not differentiate between intracellular or extracellular bacteria. To make this distinction, the addition of a detergent during homogenization should be considered and tested. In addition, unless flow cytometry analysis of free bacteria is intended to be done (most-likely with the aid of a 405 nm laser for side scatter of events smaller than eukaryotic cells) these procedures should not be considered for flow cytometry analysis of host cells as they are not intended to generate undisrupted host cell suspensions.

In summary, the procedures consist of collecting the organ, homogenization, and making serial dilutions of the homogenates. The serial dilutions are plated on BHI agar, and the plates are incubated until CFU appear and can be counted (typically 2–3 days at 26 °C). Based on the dilution used and the CFU counted, the number of CFU per tissue, or per gram of tissue if the organ was weighed, can be determined. It is important to consider that the number of CFU varies considerably from tissue to tissue as *Y. pestis* does not colonize all tissues equally. The number of CFU can also vary considerably in the same tissue depending on the time point after inoculation at which burden is assessed.

*3.3.1 Bacterial Enumeration from Soft Tissues (Lymph Nodes, Spleen, etc.) and Blood*

1. At the desired time point, humanely euthanize the mouse and collect the desired tissues. In the case of blood, this tissue can be collected using the heart puncture technique into a tube with PBS heparin. Plate on BHI agar (make dilutions for late time points, around 72 h postinoculation). In the case of lymph nodes (LNs), collect the superficial parotid LNs, which drain lymph from the ear pinna [15].
2. Place the tissues in a previously weighed small sterile plastic pouch (*see* **Note 18**).
3. Close the plastic pouch and weight it again with the tissue inside.
4. Open the plastic pouch and add 500 μL PBS (*see* **Note 19**).
5. Use a plastic cylinder or pipette as a homogenizing rod and homogenize the tissue by rolling it over the bag with pressure multiple times until the tissue is totally disrupted (*see* **Note 20**).
6. Collect 100 μL of the homogenate and plate onto a BHI plate.
7. Collect another 100 μL of the homogenate and add to a microcentrifuge tube containing 900 μL PBS. Take 100 μL of this first dilution and add to 900 μL PBS to generate a second dilution. Proceed in a similar manner if further dilutions are needed (*see* **Note 21**). Remember to mix each tube thoroughly with a vortex mixer before transferring any liquid to a new tube.
8. Plate 100 μL of each dilution on BHI agar and incubate at 26 °C for 48 h. Obtain the number of CFU per plate (*see* **Notes 21** and **22**).

*3.3.2 Bacterial Enumeration from Ear Skin*

1. At the desired time point, humanely euthanize the mouse and collect the ear (*see* **Note 23**).
2. Place the ear in a vial containing PBS (500 μL is recommended) and mince with scissors (*see* **Note 24**).
3. Add homogenizing beads and keep the vial on ice for at least 5 min (*see* **Note 25**).
4. Transfer the vial to a bead beater homogenizer (*see* **Note 26**) and homogenize following previously tested settings (*see* **Note 27**).
5. Collect 100 μL of the homogenate and plate onto a BHI agar plate with 2 μg/mL irgasan (*see* **Note 28**).
6. If further dilutions are needed (*see* **Note 29**), collect another 100 μL of the homogenate and add them to a microcentrifuge tube containing 900 μL PBS. Take 100 μL of this first dilution and add to 900 μL PBS to generate a second dilution. Proceed in a similar manner if even further dilutions are needed. Remember to mix each tube thoroughly with a vortex mixer

before transferring any liquid to a new tube. Plate as described in **step 5**.

7. Report as CFU/ear.

## 4 Notes

1. Scrape off from the area with most confluent growth, as opposed to individual colonies. Taking individual colonies increases the probability of selecting for a mutant population (e.g., those bacteria that have lost the virulence plasmid).
2. The speed at which the roller drum rotates will influence bacterial growth. However, determining what speed is optimal is an empirical process. To obtain consistent results it is very important to record the conditions that resulted in suitable growth so they can be used for every assay.
3. After incubation, the medium should look turbid. In addition, a characteristic biofilm in the form of a ring, growing on the glass tube's walls (far above the medium), should be visible.
4. While this procedure could be done using a smaller volume, larger volumes minimize error.
5. *Y. pestis* settles down in liquid culture. Therefore, cultures should be mixed thoroughly with a vortex mixer before any volume is transferred into a new tube.
6. Press the piece of double-sided tape against the tube with tweezers, making sure no wrinkles or bubbles form. Proceed in the same manner when applying the autoclave tape. Generic autoclave tape is very sturdy and sticky, and this is key to securing the animal ear in place and to prevent the formation of wrinkles as the needle penetrates into the tissue. Using tweezers instead of the finger to secure adhesive tape to the conical tube helps preserve the stickiness of the surface.
7. An inoculum of approximately 300 CFU in a volume of 2 μL (150 CFU/μL) generates highly reproducible results when assessing burden in LNs, spleen, and skin.
8. To identify the dilution predicted to have the desired bacterial concentration, the bacterial concentration of every dilution has to be determined in advance. To this end, place 2 μL of each dilution into separate microcentrifuge tubes containing 198 μL PBS. After mixing thoroughly, plate 100 μL of the mix on BHI agar plates and incubate at 26 °C for 48 h. Count the number of CFU on each plate to obtain the number of CFU/μL in each dilution. The entire procedure has to be repeated multiple times until consistent results are obtained. Being able to predict which dilution contains the desired bacterial concentration is

key because during an experiment with mice, inoculations are carried out in a blind manner. This is, the plates to quantify the inoculum will grow 48 h postinoculation and only then the actual inoculum (number of CFU in 2 μL) will be revealed.

9. It is very important to use a glass syringe when injecting small volumes as this facilitates even internal pressure in the syringe when the plunger is pressed. Plastic syringes have flexible walls that are not as responsive when injecting small volumes.
10. The use of an automatic syringe pump can be replaced with a glass syringe designed to deliver small volumes (e.g., 2–10 μL glass syringe). However, automatic syringe pumps not only guarantee very precise delivery of small volumes but also a fast and constant injection speed that can be replicated in every inoculation.
11. Press the ear pinna with the index finger making sure that the adhesive tape is completely adhered to the entire surface of the ventral side of the pinna, paying attention that there are no wrinkles in the tissue. If the ear is not perfectly flat, start all over using an unused section of the autoclave tape. It is key to use a section of tape that has not been used before because the reduced stickiness of the surface will result in unsuccessful inoculations due to the formation of wrinkles as the needle penetrates into the tissue.
12. Keep the uncapped needle always in sight and try to keep it unexposed when not in use. An open microcentrifuge tube attached to the pump can serve as a guard for the needle. This is very important to prevent accidents.
13. It is important to determine the correct settings for the syringe pump to deliver the appropriate volume (2 μL) at the appropriate speed and pressure. Because these two parameters will vary depending on the manufacturer of the device, the size of the syringe, etc., it is recommended to experiment with the equipment in advance until a visible injection wheal is successfully formed.
14. Injection is a crucial step and one cue to assess it is being done correctly is being able to visualize the bevel of the syringe very clearly through the skin. If there are doubts the bevel is being visualized clearly, most likely the needle is too deep. However, an attempt to insert the needle as shallow as possible might result in the needle escaping the tissue from the same side of the pinna it has been inserted.
15. It is recommended to practice this step on a live mouse (the wheal does not form if the animal is not alive) with PBS many times in advance before doing an experiment. Once the procedure is thought to be mastered, inoculate a test mouse and sacrifice it immediately after inoculation to homogenize and

plate the inoculated ear. This will confirm the right number of CFU were inoculated into the tissue.

16. Use the same needle for a maximum of five injections, as a blunt needle will result in suboptimal injection due to the ear wrinkling. Replacing a needle does also provide an opportunity to empty and reload the syringe, which is important because *Y. pestis* settles down over time.
17. Because it is of great importance that all mice receive the same inoculum throughout the entire experiment, it is recommended this procedure is also done before the mice are inoculated. If the system is working properly, counts should be comparable regardless when the measurements are made. Once the researcher gains more experience with the entire procedure, inoculum assessment pre-inoculation might be optional. Assessing the inoculum at the end of inoculation, however, should always be conducted.
18. Different providers sell plastic bags or pouches that are designed for these purposes.
19. Use a larger volume if you consider the size or the organ requires it.
20. This procedure can be replaced by other approaches, such as the use of an automated tissue homogenizer (e.g., bead homogenizer machine). When choosing what approach will be used, take into account the amount of heat generated by the procedure. Bead homogenizer machines produce excess heat that can compromise bacterial viability. Furthermore, consider restrictions in the BSL3 facility, as some approaches might generate more aerosols than others.
21. As previously mentioned, bacterial burden can vary drastically from tissue to tissue and from time point to time point in the same tissue. After intradermal inoculation of 300 CFU wild type (WT) *Y. pestis* bacteria should be expected to be found in the superficial parotid draining LN [15] minutes after inoculation [16], although in very low numbers. Counts increase considerably at 24 h postinoculation, the same time point at which bacteria start being detected in the spleen. At 48 h postinoculation, bacteria can be detected at high titers potentially in any tissue. For LNs, spleen, and blood, it is very likely that for later time points, bacteria will be successfully quantified only when using higher dilutions.
22. Remember that only a fraction of the original homogenate was plated, or was used for the dilutions that were plated. Thus, the dilution factor (5, if 100 μL were plated from an original volume of 500 μL) has to be taken into consideration when doing the calculations to determine CFU/organ or CFU/g of

organ. Be aware that by using 5 as a dilution factor, the volume of the tissue per se is not being taken into consideration.

23. Cut the ear from the base and make sure to be consistent between animals, always cutting at the same place.
24. Mince the ear until pieces of about 1 $mm^2$ are produced.
25. Keeping the tissue on ice is essential as homogenization with the bead beater homogenizer will increase the temperature of the sample. An overheated sample can compromise bacterial viability.
26. As opposed to soft tissues, the skin will not be disrupted unless a forcible homogenization procedure is used. Strong and controlled homogenization for small volumes is easily provided by bead beater homogenizers. Many homogenizers in the market provide excellent results. However, when selecting a particular technology consider BSL3 facility restrictions. For instance, some providers manufacture homogenizers whose homogenizing tubes are sealed, which ensures aerosols will not escape.
27. To determine optimal time of homogenization, size and material of beads (zirconia beads are a good option), etc., add a known number of CFU to a tube with the volume of PBS, number of beads, etc., to be tested and process the mix. The number of CFU obtained after processing the sample should be comparable to that of a sample that was not processed. In this way, it will be known if the procedure is detrimental to bacterial viability. The next test should be to determine if the selected conditions truly disrupt the tissue. Lastly, the procedure should be repeated in an ear immediately after inoculation. The number of CFU recovered from the ear should be comparable to that of the inoculum.
28. A large number of commensal bacteria reside on the skin and their presence will be evident if the homogenized organ is plated on agar plates with no selection. For this reason, when using WT *Y. pestis* strains, it is necessary to use BHI agar plates with irgasan, a compound to which *Y. pestis* is naturally resistant and most skin commensals are not. For genetically engineered strains with a resistance marker the use of irgasan might not be necessary, provided plating is done on media with the appropriate antibiotics.
29. For inoculations in WT C57BL/6 mice with WT *Y. pestis* CO92, do not expect bacterial numbers to increase over time. The median number of CFU recovered from this tissue should be similar to that of the inoculum. However, assessments of burden in the ear after systemic dissemination (around 24 h postinoculation with an inoculum of approximately 300 CFU) will reveal not only bacterial counts in the skin per se but also in the blood circulating in the tissue. A

useful control is to determine burden in both ears. Bacteria should be absent in the ear that is not inoculated unless systemic dissemination has already taken place.

## Acknowledgments

The author would like to thank Virginia L. Miller for kindly reviewing and editing this work.

### References

1. Bacot AW, Martin CJ (1914) Observations on the mechanism of the transmission of plague by fleas. J Hyg (Lond) 13:423–439
2. Hinnebusch BJ (2005) The evolution of flea-borne transmission in *Yersinia pestis*. Curr Issues Mol Biol 7:197–212
3. Lorange EA, Race BL, Sebbane F, Hinnebusch BJ (2005) Poor vector competence of fleas and the evolution of hypervirulence in *Yersinia pestis*. J Infect Dis 191:1907–1912
4. Demeure CE, Blanchet C, Fitting C, Fayolle C, Khun H, Szatanik M, Milon G, Panthier J-J, Jaubert J, Montagutelli X, Huerre M, Cavaillon J-M, Carniel E (2012) Early systemic bacterial dissemination and a rapid innate immune response characterize genetic resistance to plague of SEG mice. J Infect Dis 205:134–143
5. Cathelyn JS, Crosby SD, Lathem WW, Goldman WE, Miller VL (2006) RovA, a global regulator of *Yersinia pestis*, specifically required for bubonic plague. Proc Natl Acad Sci U S A 103:13514–13519
6. Oyston PC, Dorrell N, Williams K, Li SR, Green M, Titball RW, Wren BW (2000) The response regulator PhoP is important for survival under conditions of macrophage-induced stress and virulence in *Yersinia pestis*. Infect Immun 68:3419–3425
7. Guinet F, Carniel E (2003) A technique of intradermal injection of *Yersinia* to study *Y. pestis* physiopathology. Adv Exp Med Biol 529:73–78
8. Sebbane F, Jarrett CO, Gardner D, Long D, Hinnebusch BJ (2006) Role of the *Yersinia pestis* plasminogen activator in the incidence of distinct septicemic and bubonic forms of flea-borne plague. Proc Natl Acad Sci U S A 103:5526–5530
9. Chong SZ, Evrard M, Ng LG (2013) Lights, camera, and action: vertebrate skin sets the stage for immune cell interaction with arthropod-vectored pathogens. Front Immunol 4:286
10. Gonzalez RJ, Weening EH, Lane MC, Miller VL (2015) Comparison of models for bubonic plague reveals unique pathogen adaptations to the dermis. Infect Immun 83(7):2855–2861. https://doi.org/10.1128/IAI.00140-15
11. Teunissen MBM, Haniffa M, Collin MP (2012) Insight into the immunobiology of human skin and functional specialization of skin dendritic cell subsets to innovate intradermal vaccination design. Curr Top Microbiol Immunol 351:25–76
12. Combadiere B, Liard C (2011) Transcutaneous and intradermal vaccination. Hum Vaccin 7:811–827
13. Shayan R, Achen MG, Stacker SA (2006) Lymphatic vessels in cancer metastasis: bridging the gaps. Carcinogenesis 27:1729–1738
14. Skobe M, Detmar M (2000) Structure, function, and molecular control of the skin lymphatic system. J Investig Dermatol Symp Proc 5:14–19
15. Van den Broeck W, Derore A, Simoens P (2006) Anatomy and nomenclature of murine lymph nodes: descriptive study and nomenclatory standardization in BALB/cAnNCrl mice. J Immunol Methods 312:12–19
16. Gonzalez RJ, Lane MC, Wagner NJ, Weening EH, Miller VL (2015) Dissemination of a highly virulent pathogen: tracking the early events that define infection. PLoS Pathog 11: e1004587

# Chapter 2

# Intranasal Inoculation of Mice with *Yersinia pestis* and Processing of Pulmonary Tissue for Analysis

**Roger D. Pechous**

## Abstract

Pneumonic plague is a rapidly progressing and highly lethal pneumonia caused by pulmonary infection with *Yersinia pestis.* Disease is marked by the rapid replication of bacteria in the lungs in the absence of symptoms, followed by the abrupt onset of a highly lethal inflammatory response. A murine intranasal infection model has been key to characterizing the progression of disease. Mice are a natural *Y. pestis* host, and murine disease closely mirrors what is seen during human infection. Intranasal inoculation of mice with fully virulent *Y. pestis* strains allows for the detailed analysis of key bacterial and host factors that define disease progression. In this chapter I describe a method for intranasal inoculation of mice with *Y. pestis*, as well as techniques for processing lung tissue for analysis. These include protocols for isolating whole lungs and lavage fluid for measure of bacterial burden, transcriptomics, cytokine/chemokine expression, and flow cytometry. These techniques can be used to evaluate disease parameters of interest during typical infection, infection with bacterial mutants, or infection in the presence of pharmacological agents aimed at targeting specific host or bacterial factors. Combining a highly relevant murine infection model with these techniques provides a powerful platform for fully evaluating the progression of pneumonic plague.

**Key words** Plague, *Yersinia*, Pneumonic plague, Intranasal infection, Pulmonary inflammation, Pneumonia, Bronchoalveolar lavage

## 1 Introduction

Inhalation of respiratory droplets containing *Yersinia pestis* results in primary pneumonic plague, a rapidly progressing and highly lethal pulmonary infection. Pneumonic plague is the most severe manifestation of *Y. pestis* infection, and is almost always lethal in the absence of treatment [1–3]. A mouse intranasal model of *Y. pestis* infection has been critical to characterizing the progression of pneumonic plague [4–7]. Mice are a natural host for *Y. pestis*, and disease in mice closely resembles human infection [8, 9]. The murine intranasal infection model is defined by an initial "preinflammatory" phase lasting roughly 36 h that sees a rapid period of bacterial proliferation in the absence of disease symptoms or signs of inflammation [4, 6]. This is followed by the abrupt switch to an intense

Viveka Vadyvaloo and Matthew B. Lawrenz (eds.), *Pathogenic Yersinia: Methods and Protocols*, Methods in Molecular Biology, vol. 2010, https://doi.org/10.1007/978-1-4939-9541-7_2, 

"proinflammatory" disease phase characterized by the expression of a proinflammatory cytokine storm, the formation of neutrophil-rich pulmonary lesions, and the onset of symptoms [4, 6]. Progression of the proinflammatory disease phase occurs rapidly, and results in severe pulmonary necrosis, destruction of alveolar architecture, and consolidation of entire lobes of the lung immediately prior to death at roughly 72 h post-infection. Disease progression highlights two key "failures" of the innate immune system: (1) the failure to initiate an appropriate innate immune response to control infection and (2) dysregulation of innate immune mechanisms that culminate in an intense and highly damaging proinflammatory disease phase that ultimately proves lethal. These characteristics are common to many models of pneumonia caused by a number of both viral and bacterial pathogens, and contribute to difficulty treating late-stage severe pulmonary infection [10, 11].

The availability of a robust, reproducible, and relevant mouse model of infection allows for the detailed characterization of key bacterial and host factors that contribute to disease. In the methods described here, I outline a simple protocol for intranasal inoculation of mice with *Y. pestis*, and highlight standard techniques for processing pulmonary tissue for analysis. These include lung dissection and homogenization, as well as lung perfusion for elimination of blood from the pulmonary vasculature, thus allowing for isolation of the interstitial and alveolar compartments. I also describe protocols for lung fixation and collection of the bronchoalveolar lavage fluid (BALF), both of which require catheterization of the trachea and inflation of the lungs. This is a powerful technique, as it can be used to generate inflated lung samples for histopathological analysis, single-cell suspensions for flow cytometry and/or microscopic analysis, and BALF samples for the measure of cytokines/chemokines, host cell types, and markers of pulmonary damage [12–15]. The procedures described here can be used to measure multiple disease parameters simultaneously, allowing for a complete and detailed evaluation of the progression of pneumonic plague. Further, using these techniques to analyze host responses to various bacterial mutants and under a variety of experimental timings and conditions allows for a comprehensive evaluation of infection for the purpose of identifying key bacterial and host mediators of disease.

## 2 Materials

### 2.1 Culture Preparation and Intranasal Inoculation

1. 1× phosphate-buffered saline (PBS).
2. Ketamine–xylazine cocktail (or a preferred anesthetic).
3. Brain–heart infusion (BHI) broth and agar.
4. Culture tubes.

5. Calcium chloride.
6. P20 pipette tips.
7. Mice, 6–8 weeks old (*see* **Note 1**).

### 2.2 Lung Isolation and Homogenization

1. 1× phosphate-buffered saline (PBS).
2. TRIzol reagent.
3. 50 ml conical tubes, with a hole in the lid slightly larger than the diameter of Tissue-Tearor probe.
4. Tissue-Tearor/homogenizer (*see* **Note 2**).
5. Mice necropsy tools: one pair of small to medium sized dissection scissors, one pair of straight forceps.
6. Pin board, Styrofoam square, platform.
7. Syringe needles (any size).
8. 70% ethanol.
9. Sterile water.
10. 0.2 μm syringe filter.

### 2.3 Lung Perfusion

1. 1× phosphate-buffered saline (PBS).
2. Pin board, Styrofoam square, platform.
3. Mice necropsy tools: one pair of dissection scissors, one pair of straight forceps.
4. 5 ml syringe.
5. 25-gauge syringe needles.
6. Gauze pads.

### 2.4 BALF Collection

1. 1× phosphate-buffered saline (PBS).
2. Ice bucket/ice.
3. 3-0 Silk surgical suture cut into 4- to 6-in. pieces.
4. 22-gauge, 1-in. catheter.
5. Gauze pads.
6. Mice necropsy tools: one pair of small microdissection scissors, one pair of straight forceps, one pair of curved precision tip forceps, and one pair of pointed straight forceps.
7. Pin board, Styrofoam square, pinning platform.
8. Syringe needles (any size).
9. 15 or 50 ml conical tubes.

### 2.5 Lung Inflation for Fixation

1. 10% formalin.
2. 3-0 Silk surgical suture cut into 4- to 6-in. pieces.

3. 22-gauge, 1-in. catheter.
4. Mice necropsy tools: one pair of small microdissection scissors, one pair of straight forceps, one pair of curved precision tip forceps, one pair of pointed straight forceps.
5. Pin board, Styrofoam square, pinning platform.
6. Syringe needles (any size).
7. 50 ml conical tubes.

### 2.6 Generating a Single-Cell Suspension for Flow Cytometry

1. 3-0 Silk surgical suture cut into 4 to 6-in. pieces.
2. 22-gauge, 1-in. catheter.
3. Mice necropsy tools: one pair of small microdissection scissors, two pair of straight forceps, one pair of curved precision tip forceps, one pair of pointed straight forceps.
4. Pin board, Styrofoam square, pinning platform.
5. Syringe needles (any size).
6. Dispase I solution: 5 U/ml Dispase 1 in 1× PBS.
7. Tissue disruption solution: 1× PBS with 100 μg/ml DNAse I.
8. Cell Strainer, 80 or 100 μm.
9. Petri plates.
10. 50 ml conical tubes.
11. RBC lysis buffer: 0.15 M $NH_4Cl$, 10 mM $KHCO_3$, 0.1 mM EDTA.
12. FACS buffer: 1× PBS, 2% fetal bovine serum.
13. Fc blocking antibody: 2.4G2 hybridoma supernatant.
14. 2% Formalin.

## 3 Methods

### 3.1 Culture Preparation and Intranasal Inoculation

1. Grow *Y. pestis* strains on BHI agar at 26 °C for 2 days.
2. Inoculate 2 ml BHI broth with multiple colonies of *Y. pestis* (*see* **Note 3**) and grow for 6–12 h at 26 °C in roller drum or with constant shaking at 225 rpm.
3. Dilute turbid culture to an optical density at 620 nm ($OD_{620}$) of 0.05–0.1 in BHI supplemented with 2.5 mM $CaCl_2$ and grow for 12–16 h at 37 °C with constant shaking (225 rpm).
4. Centrifuge 1 ml culture at 18,000 × *g* for 1 min and resuspend in 1 ml 1× PBS.
5. Anesthetize mice via intraperitoneal injection with ketamine–xylazine.

6. Take $OD_{620}$ reading of culture in1× PBS, dilute to desired concentration (colony-forming units per 20 μl volume) and inoculate 20 μl directly into the nares of anesthetized mice (*see* **Note 4**). When inoculating mice, scruff anesthetized mice behind the neck and hold at a roughly 45-degree angle with the head in a slightly elevated position. Putting the pipette tip as close as possible to the nares, slowly depress the plunger in a single motion as the animal breathes and inhales the inoculum.
7. Place mice on their side in their cage to allow for recovery from anesthesia.

### 3.2 Lung Isolation and Homogenization

1. Euthanize mouse per institutional guidelines (*see* **Note 5**) and pin to a platform using syringe needles in each foot with the ventral side of the mouse facing up.
2. Soak body of mouse with 70% ethanol.
3. Using forceps, lift skin of lower chest and cut through skin and fascia with scissors, opening a hole that allows access to the peritoneal cavity and chest. Cut down the midsagittal line toward the posterior end of the animal exposing the peritoneal cavity (Fig. 1).
4. Moving the liver aside, carefully cut the diaphragm to gain access to the chest cavity (Fig. 1). If desired, cut up the sides of the chest cavity and across sternum to remove chest plate.
5. Use a wide grip with forceps to grab as much of the lung tissue as possible, and gently lift while snipping trachea, tissue connecting the lungs to the diaphragm, and surrounding tissue to release the lungs from the chest cavity (*see* **Note 6**).
6. Place lungs into 50 ml conical tube containing 1 ml 1× PBS for enumeration of bacterial burden or measure of cytokines/chemokines, or 1 ml TRIzol or desired RNA isolation reagent.
7. Using lid from conical tube with hole in center, insert probe from Tissue-Tearor into the hole in the lid, and screw the lid onto the conical containing the organ to be homogenized (*see* **Note 7**). Homogenize on medium to high, moving up and down slightly until the organ is homogenized into a uniform suspension.
8. For RNA isolation, proceed with reagent instructions per manufacturer's protocol. For enumeration of bacteria, perform serial dilutions and plate counts per standard microbiological procedures. For measure of pulmonary cytokine/chemokine secretion, centrifuge lung homogenate at 18,000 × *g* for 5 min at 4 °C and collect supernatant for analysis using ELISA or bead array platforms. The supernatant can be passed through a syringe filter if necessary to eliminate bacteria and debris.

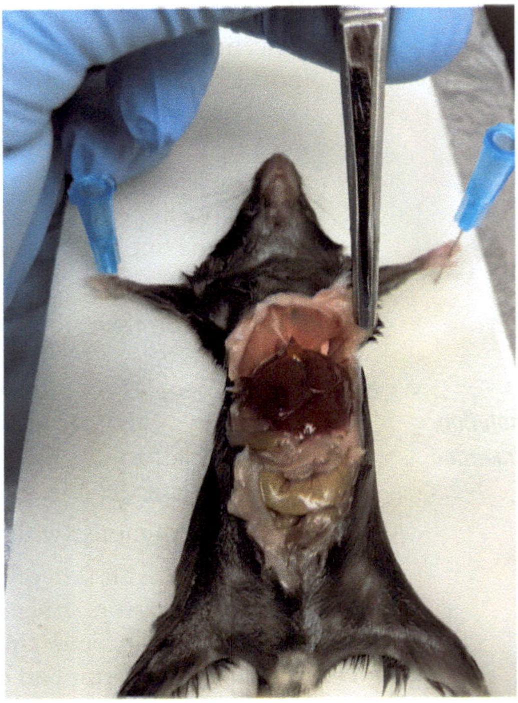

**Fig. 1** Mouse dissection for processing of pulmonary tissue. Carefully cut down the midsagittal line of the mouse exposing the diaphragm and peritoneal cavity. Carefully nick the diaphragm to expose the chest cavity

### *3.3 Lung Perfusion*

1. Euthanize mouse per institutional guidelines and pin to a platform using syringe needles in each foot with the ventral side of the mouse facing up.
2. Soak body of the mouse with 70% ethanol.
3. Using forceps, pinch skin of lower chest and cut through skin and fascia with scissors. Cut down the midsagittal line toward the posterior end of the animal exposing the peritoneal cavity.
4. Cut vasculature leading to the kidney just above the organ to allow for draining of blood, and place two gauze pads to soak.
5. Moving the liver aside, carefully nick the diaphragm and cut to gain access to the chest cavity (Fig. 1).
6. Immobilize the heart using forceps and insert the syringe needle into the right ventricle immediately to the left of the apex/tip of the heart. Slowly depress the syringe, perfusing with up to 5 ml 1× PBS. The PBS should begin to drain out of the cut vasculature leading to the kidney. Lungs should begin to turn very light pink to white.

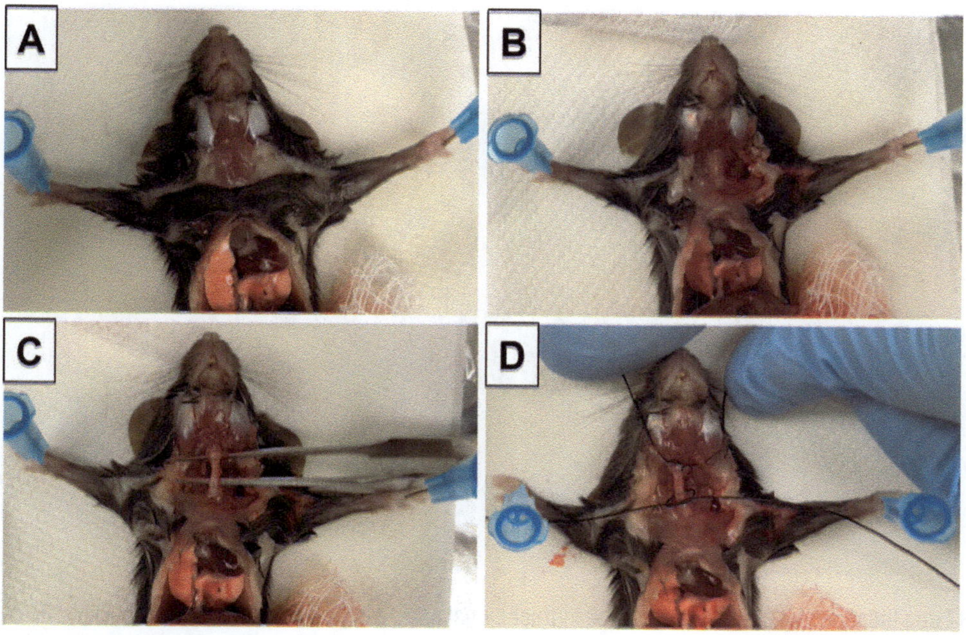


**Fig. 2** Exposing the trachea for catheterization. (**a**) Carefully cut away skin to expose underlying tissue. (**b**) Remove fatty tissue, salivary glands, and muscle tissue to isolate the trachea and tightly associated smooth muscle. (**c**) Remove tissue immediately surrounding trachea, allowing for complete access and visibility of tracheal rings. (**d**) Tie silk thread loosely around trachea using forceps and slide knot down toward chest cavity. Tie second silk thread around trachea closer to the head of the animal, slide as far up toward the larynx and tighten. Tug slightly on "top" thread closes to head to extend trachea for ease of manipulation

### 3.4 BALF Collection

1. With chest and peritoneal cavity exposed, lift the liver and cut the hepatic portal vein exiting the liver toward the posterior of the animal, and cover with gauze pad to bleed (*see* **Note 8**).
2. Lift skin on ventral side roughly midway down the neck with forceps and cut, exposing anatomy of the neck (Fig. 2a). Using forceps and scissors, remove fatty tissue and salivary glands, exposing trachea and surrounding smooth muscle (Fig. 2b). Carefully remove thin muscle/tissue surrounding trachea, using pointed forceps or scissors to remove tissue under and around trachea allowing for full exposure (Fig. 2c).
3. Using forceps, tie a loose knot of silk suture around the trachea and slide as far down toward the chest cavity as possible, leaving untightened. Tie a second loose knot of silk and bring as close to the top (toward head) of trachea as possible and tighten (Fig. 2d).
4. Carefully pull the tightened "top" suture proximal to the animal's head to stretch out the trachea (*see* **Note 9**) (Fig. 2d). Using a syringe needle, scrape or poke small opening on front of trachea between tracheal rings on the upper half of the trachea just below the larynx (*see* **Note 9**). Carefully insert the catheter into the small opening in the trachea and slide up to

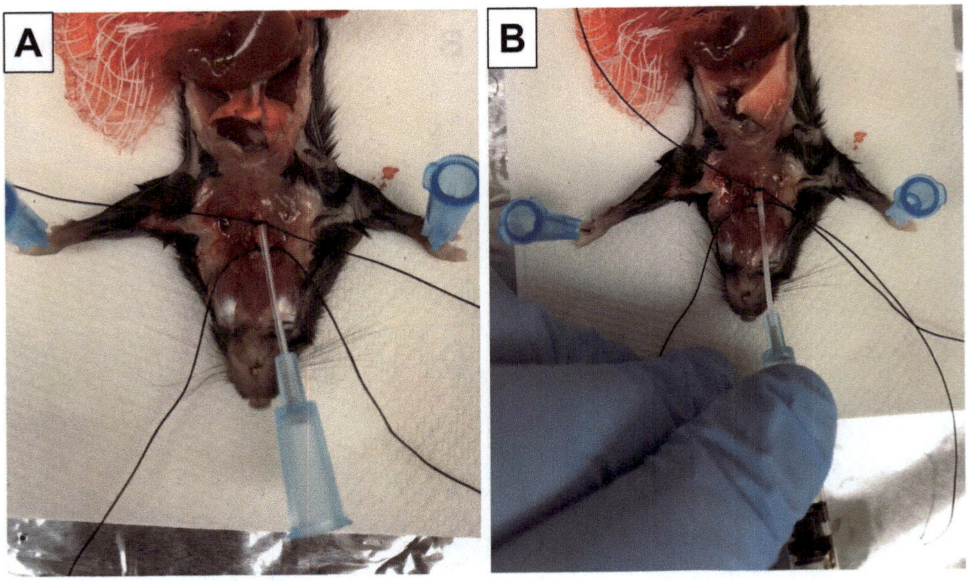


**Fig. 3** Lung inflation. (**a**) Create a slight nick in trachea using a syringe needle, and carefully slide the catheter into the trachea, through the loose silk knot. Tighten the silk knot around the trachea to stabilize. (**b**) Attach syringe containing fluid to inflate lung at port of the catheter and slowly depress plunger until lung is inflated with roughly 900 μl

roughly one-fourth the length of the catheter (enough to secure with thread, but before reaching bronchi) into the trachea and through the loosely tied silk suture (*see* **Note 9**). Tighten the suture to secure the catheter within the trachea (Fig. 3a).

5. To collect the BALF, attach the base of a 1 ml syringe (without needle) containing 1 ml 1× PBS to the catheter and slowly inject into the mouse lung, watching for leaks as the lung inflates (Fig. 3b). Hold the catheter port steady to stabilize. Slowly withdraw the fluid back into the syringe (*see* **Note 10**) and add the fluid to a 15 ml or 50 ml conical tube on ice. Repeat twice, or until fluid is clear for animals in the later stages of infection.
6. Centrifuge BALF at 500 × *g* for 5 min at 4 °C. For evaluating cytokines or enumerating bacterial burden, resuspend cell pellet in desired volume of 1× PBS or flow buffer. For flow cytometry, proceed with red blood cell lysis in Subheading 3.6, **step 4**.

### 3.5 Lung Inflation for Fixation

1. To inflate lungs for fixation, follow procedure outlined above (Subheading 3.4), and fill lungs with 1 ml 10% formalin. When lungs are visibly inflated, remove syringe, slide catheter out and tighten suture around trachea.

2. To remove inflated lungs (heart is still attached), grab excess suture with forceps and lift. Cut trachea above the suture and while lifting trachea/lungs, carefully snip under and around lungs with scissors until intact lungs with attached trachea and heart are free from the rest of the animal.
3. Place lungs in a 50 ml conical tube containing 15 ml of 10% formalin and allow at least 1 h at room temperature or 4 °C for complete penetration. Lungs are now ready for embedding, slicing, staining, and histopathological analysis (*see* **Note 11**).

### 3.6 Generating a Single-Cell Suspension for Flow Cytometry

1. To generate a single-cell suspension, follow procedure outlined above (Subheading 3.4), and fill lungs with 1 ml Dispase I solution. When lungs are visibly inflated, remove syringe, slide catheter out, and tighten lower suture around trachea. Grab excess suture with forceps and lift gently, cutting the trachea above the suture and surrounding tissue to free lungs (and attached heart) from the chest cavity.
2. Add inflated lungs to a 50 ml conical tube containing 15 ml Dispase I solution and incubate at room temperature for 25 min.
3. Place lungs in a petri plate with 10 ml tissue disruption solution and remove the heart with scissors. Using forceps, grip trachea/connective tissue firmly and use a second set of forceps to disrupt tissue and generate a suspension using a repeated clamping motion (*see* **Note 12**). Gently swirl for 1 min and pass through a 100 μm nylon cell strainer into a 50 ml conical tube.
4. Spin filtered suspension at 500 × *g* for 5 min at 4 °C and resuspend cell pellet in 1 ml red blood cell (RBC) lysis solution. After incubation at room temperature for 2 min, add 9 ml 1× PBS, and spin at 500 × *g* for 5 min (*see* **Note 13**).
5. If staining for flow cytometry, resuspend cell pellet in 100 μl 2.4G2 hybridoma supernatant (*see* **Note 14**) and incubate at 4 °C for 30 min. Spin at 500 × *g* for 5 min at 4 °C and resuspend cell pellet in 100–500 μl FACS buffer containing desired fluorescently labeled antibodies. Incubate at 4 °C for 30–60 min, followed by centrifugation at 500 × *g* for 5 min. Resuspend in a desired volume of FACS buffer (*see* **Note 15**).

## 4 Notes

1. We typically use 6–8-week-old female C57Bl/6 mice. We have also examined infection in 6–8-week-old male mice and in male and female mice up to 20 weeks old and saw no appreciable difference in disease progression.

2. If possible, a battery-operated device is preferred, as it is typically lighter and easier to handle. Any type of tissue homogenizer will work, as long as the organ can be brought to a uniform suspension.
3. We typically collect a light "smear" of single colonies to inoculate a culture when using fully virulent strains as opposed to using a single colony per standard microbiological techniques. This is because *Y. pestis* can spontaneously lose the pCD1 virulence plasmid or the *pgm* virulence locus resulting in attenuation of the strain.
4. For enumerating bacterial culture concentration to calculate an inoculating dose, we use $2 \times 10^8$ CFU/ml for an $OD_{620}$ of 1.0. Alternatively, the inoculum can be delivered dropwise, taking a few seconds to allow the animal to breathe between roughly every 5 μl of inoculum.
5. Our preferred method of euthanasia is using sodium pentobarbital or another agent delivered via intraperitoneal injection. Most of the described techniques generate sample for downstream analysis of pulmonary inflammation and/or physiology, and it is possible that inhalation of anesthetics may alter results. We therefore recommend intraperitoneal injection of anesthetic when possible.
6. The lungs can be "slippery" and difficult to maintain a grip using forceps without appreciable tearing or breaking into pieces. Since the lungs are to be homogenized for bacterial enumeration or RNA isolation, it is fine to simply collect all the lung tissue pieces that are left in the chest cavity if the tissue does tear and homogenize together.
7. We typically use scissors to round out a hole in the top of one or two conical tube lids and use a single lid for an entire experiment or a set of samples. As samples likely contain *Y. pestis*, this is to avoid excessive aerosolization of the organism. Alternatively, tissue can be placed in a small tissue bag with 1 ml of 1× PBS and a roller used to flatten and homogenize tissue.
8. Bleeding the animal in this manner makes dissection of the throat much easier and cleaner, as otherwise during dissection blood tends to pool in the neck. If excess blood does pool in the neck, simply soak it up with gauze, as it will interfere with tracheal cannulation and dissection of the surrounding tissue.
9. Using the top suture to slightly stretch the trachea makes it easier to see between the cartilage rings and tightens the trachea allowing for puncturing with a syringe needle. It is best to use a slight scraping with the tip of the needle to generate a very small hole, which will appear black against the rest of the tissue. Be careful not to make the hole too large, as if the trachea is severed it will retract into the lungs and inflation of the lungs

becomes very difficult. Carefully insert the catheter into the hole, first at a roughly 90° angle, then slowly slide the catheter tip forward into the trachea.

10. Try to remove the lavage fluid in one slow and continuous motion. It is very difficult to remove the entire volume initially injected into the lungs, and it is acceptable if a small volume (100–200 μl) is left behind. Inevitably the catheter tip will touch tracheal tissue and suction will block the withdrawal of BALF, resulting in drawing up of air and lungs remaining inflated. Slowly and gently move the catheter around, moving forward and backward, being careful not to move too far forward that tissue is punctured or pulling too far backward such that the catheter is pulled out of the trachea. While adjusting the catheter pull the syringe plunger slowly in and out until liquid begins to be drawn up again. It may be necessary to inject a small additional volume of PBS and retry withdrawing BALF in one continuous motion if unable to "unstick" the catheter.

11. If it is necessary to remove a sample from biosafety level 3 containment, it is suggested that a duplicate sample is generated and the lung homogenized as in Subheading 3.2. A portion of the sample (at least 10% of the total sample volume or 500 μl sample, whichever is smaller) is diluted into 2 ml BHI and incubated at 26 °C for at least 6 h prior to being plated on BHI and incubated a minimum of 3 days at 26 °C to confirm the absence of viable agent. Alternatively, a 500 μl volume of the formalin the tissue is incubated in can be tested as described.

12. For best results, grip the trachea and connective tissue tightly, and use the second set of forceps in a repeated clamping motion to effectively "smash" the lung tissue. If digestion was not complete, it is possible to pull or tease apart tissue that is difficult to get into suspension, though it is suggested that those small pieces of tissue are continuously teased apart and clamped until they are as small as possible. If done properly, the tissue dissipates into a suspension quite easily, and you will be left with a cell suspension as well as a continuous trachea/airway piece of tissue that can be discarded. This typically takes 5 min or less.

13. RBC lysis can be performed for up to 5 min at room temperature. After centrifugation, there should be a very minimal amount of visible blood. If an initial RBC lysis step did not result in complete RBC lysis, and significant amounts of blood are visible, this step can be repeated.

14. Though incubation with Fc block is common to most protocols, it is of note that we saw little to no difference in staining in

the presence or absence of Fc block, and typically list this step as optional.

15. If fixation is necessary for removal from Biosafety level 2 or 3 containment for analysis, spin sample at 500 × *g* for an additional 5 min at 4 °C, and resuspend in 100 µl of at least 2% formalin. Incubate for a minimum of 15 min at room temperature, spin at 500 × *g* for 5 min, and resuspend in the desired volume of FACS buffer.

## Funding

Funding was provided by COBRE NIH/NIGMS grant P20-GM103625 provided through the UAMS Center for Microbial Pathogenesis and Host Inflammatory Responses.

### References

1. Perry RD, Fetherston JD (1997) *Yersinia pestis*–etiologic agent of plague. Clin Microbiol Rev 10:35–66
2. Krishna G, Chitkara RK (2003) Pneumonic plague. Semin Respir Infect 18:159–167
3. Inglesby TV, Dennis DT, Henderson DA et al (2000) Plague as a biological weapon: medical and public health management. Working group on civilian biodefense. JAMA 283 (17):2281–2290
4. Lathem WW, Crosby SD, Miller VL et al (2005) Progression of primary pneumonic plague: a mouse model of infection, pathology, and bacterial transcriptional activity. Proc Natl Acad Sci U S A 102:17786–17791
5. Lathem WW, Price PA, Miller VL et al (2007) A plasminogen-activating protease specifically controls the development of primary pneumonic plague. Science 315:509–513
6. Bubeck SS, Cantwell AM, Dube PH (2007) Delayed inflammatory response to primary pneumonic plague occurs in both outbred and inbred mice. Infect Immun 75:697–705
7. Price PA, Jin J, Goldman WE (2012) Pulmonary infection by *Yersinia pestis* rapidly establishes a permissive environment for microbial proliferation. Proc Natl Acad Sci 109:3083–3088
8. Guarner J, Shieh W-J, Greer PW et al (2002) Immunohistochemical detection of *Yersinia pestis* in formalin-fixed, paraffin-embedded tissue. Am J Clin Pathol 117:205–209
9. Doll JM, Zeitz PS, Ettestad P et al (1994) Cat-transmitted fatal pneumonic plague in a person who traveled from Colorado to Arizona. Am J Trop Med Hyg 51:109–114
10. Bordon J, Aliberti S, Fernandez-Botran R et al (2013) Understanding the roles of cytokines and neutrophil activity and neutrophil apoptosis in the protective versus deleterious inflammatory response in pneumonia. Int J Infect Dis 17:e76–e83
11. Restrepo MI, Anzueto A (2009) Severe community-acquired pneumonia. Infect Dis Clin N Am 23:503–520
12. Pechous RD, Sivaraman V, Price PA et al (2013) Early host cell targets of *Yersinia pestis* during primary pneumonic plague. PLoS Pathog 9:e1003679
13. Pechous RD, Broberg CA, Stasulli NM et al (2015) In vivo transcriptional profiling of *Yersinia pestis* reveals a novel bacterial mediator of pulmonary inflammation. MBio 6: e02302–e02314
14. Sahoo M, Del Barrio L, Miller MA et al (2014) Neutrophil elastase causes tissue damage that decreases host tolerance to lung infection with Burkholderia species. PLoS Pathog 10: e1004327
15. Matute-Bello G, Downey G, Moore BB et al (2011) An official American thoracic society workshop report: features and measurements of experimental acute lung injury in animals. Am J Respir Cell Mol Biol 44:725–738

# Chapter 3

# Standardized Method for Aerosol Challenge of Rodents with *Yersinia pestis* for Modeling Primary Pneumonic Plague

**Paul E. Anderson, Rachel M. Olson, Joshua L. Willix, and Deborah M. Anderson**

## Abstract

Primary pneumonic plague occurs when *Yersinia pestis* is inhaled into the lower respiratory tract where it invades the alveoli and grows. Rapid bacterial growth eventually elicits a neutrophilic inflammatory response that is ineffective and damaging, leading to accelerated progression of disease. In the laboratory, modeling of primary pneumonic plague can be accomplished by instillation of bacterial culture in the nares of anesthetized mice and rats. Although primary pneumonic plague can develop from this method, variability in dosing and side effects of anesthesia can complicate data interpretation. In contrast, aerosol challenge models allow for well-controlled studies of pneumonic plague with minimal experimental bias and unwanted side effects. For these reasons, antibiotic testing and the licensing of new treatments depend on efficacy data generated from aerosol delivery of *Y. pestis* in order to more accurately model transmission and the early stages of human pneumonic plague. In order to meet this need, we have extensively characterized pneumonic plague in mice and rats challenged by nose-only exposure to *Yersinia pestis*. With this approach, simultaneous challenge of large cohorts of animals, gently restrained and not anesthetized, assures safe, well-controlled, unbiased, and uniform infection. In this chapter, we present a standardized method for reproducible aerosol delivery of wild-type *Y. pestis* to rodents for experimental models of primary pneumonic plague.

**Key words** *Yersinia pestis*, Pneumonic plague, Aerosol, Animal model

## 1 Introduction

Primary pneumonic plague develops in humans and other mammals when *Y. pestis* aerosols are inhaled. Following inhalation, deposition in the lower lung occurs and results in focal bacterial growth in the bronchial alveoli followed by vascular dissemination. In the typical research laboratory, primary pneumonic plague is studied in the mouse model following intranasal (IN) inoculation of anesthetized animals. This method can be problematic as the inoculum can be retained in the upper respiratory tract leading to a

Viveka Vadyvaloo and Matthew B. Lawrenz (eds.), *Pathogenic Yersinia: Methods and Protocols*, Methods in Molecular Biology, vol. 2010, https://doi.org/10.1007/978-1-4939-9541-7_3, © Springer Science+Business Media, LLC, part of Springer Nature 2019

high percentage of animals developing secondary septicemic plague with only a low degree of primary lung infection [1]. For highly virulent organisms like *Y. pestis*, systemic spread from the upper respiratory tract leads to the rapid development of secondary septicemic plague, a disease that progresses to lethality with kinetics that resemble the pneumonic form despite very different pathology and cause of death. Furthermore, variability within groups can occur with the IN model because individual mice are challenged instead of whole cohorts. Other problems associated with IN delivery include variations introduced by different methods of anesthesia and the downstream impact on cilia and the inflammatory response. For example, isoflurane depresses cilia and can induce anti-inflammatory IL10 [2, 3]. Ketamine–xylazine is used for a longer-lasting sedation but presents a greater risk to the animal due to the potential for hypothermia, hypoglycemia, and/or respiratory or cardiovascular problems that can be life threatening [4].

Aerosol challenges, however, can be performed simultaneously on groups of rodents without anesthesia in a highly controlled environment that induces minimal stress [5]. Inhalation exposure systems for modeling infectious disease in rodents are used in biodefense applications and tuberculosis research as well. Numerous methods for aerosol generation and delivery have been developed, along with accessories and options that customize the equipment according to research needs, available facilities, and biohazard risk. Inhalation exposure chambers for biodefense research are highly individualized, with virtually no two systems identical. A strong understanding of the equipment, pathogen, facility, and model is necessary to develop protocols for highly reproducible infection by microbial aerosols. Below we present methods used in our laboratory for *Yersinia pestis* and discuss considerations and alternatives in Subheading 4.

## 2 Materials

### 2.1 Preparation of Challenge Material

1. Heart infusion broth (HIB).
2. Heart infusion agar (HIA).
3. Congo Red Plates: HIA supplemented with 0.2% Galactose and 0.005% Congo Red.
4. 20% galactose in $H_2O$, filter sterilized.
5. 0.5% Congo Red in $H_2O$, filter sterilized.
6. 2× Bacteriological freezing media: 10% bovine serum albumin (BSA), 10% monosodium glutamate (MSG) in $H_2O$, filter sterilized (*see* **Note 1**)
7. 1 M calcium chloride ($CaCl_2$) in $H_2O$, autoclaved.
8. Disposable 250 mL vented flasks.
9. Sterile phosphate-buffered saline (PBS), pH 7.4.

10. Disposable cuvettes and visible light spectrophotometer.
11. 50 mL centrifugation tubes.

### 2.2 Aerosol Challenge

1. Class II or Class III Biosafety Cabinet, with sealed penetrations for operation of air compressors and vacuum pumps.
2. Sparging liquid aerosol generator (SLAG) nebulizer (*see* **Note 2**), with 0.5–2.0 μm pore size metallic frit (*see* **Note 3**).
3. Air compressor.
4. Tubing and inline HEPA filters.
5. Peristaltic pump.
6. Teflon (or glass) impingers (*see* **Note 4**).
7. Antifoam (*see* **Note 5**).
8. Appropriately sized animal restraint devices with pushers/plungers for nose-only exposure chamber (*see* **Note 6**).
9. Silicon o-ring lubricant.
10. Nose-only exposure chamber, with negative tower pressure.
11. Vacuum pump.
12. Mass flow controllers (*see* **Note** 7).
13. Particle sizing spectrometer (optional).
14. Thermometer and hygrometer (optional).
15. 70% ethanol.
16. 10% bleach (0.5% sodium hypochlorite).
17. Vaporous hydrogen peroxide generator.
18. 30% hydrogen peroxide ($H_2O_2$).

## 3 Methods

### 3.1 Preparation of Challenge Material (See Note 8)

1. To prepare working stocks, sufficient for completing a set of studies, streak *Y. pestis* from frozen stock on Congo Red plates to verify retention of the pigmentation locus and purity of starting material [6].
2. Incubate plate for 48–72 h at 26 °C.
3. Use isolated, pigmented colonies for inoculation of HIB containing 2.5 mM $CaCl_2$. For preparation of a batch of working stocks, we typically inoculate 40 mL of media and grow for approximately 24 h at 26 °C with aeration (120 rpm).
4. Measure $OD_{600}$ and dilute to desired concentration in HIB or 1× Bacteriological freezing media to make working stocks (*see* **Note 9**). Prepare a sufficient number of working stocks to complete a study set.
5. Working stocks are stored at −80 °C in 1× bacteriological freezing media.

6. To prepare challenge material, completely thaw one frozen vial of the working stock and dilute into HIB + 2.5 mM $CaCl_2$ in a 250 mL flask and grow for 22–24 h at 37 °C with aeration (*see* **Note 10**).
7. Measure $OD_{600}$. Centrifuge at 5000 × *g* for 15 min (*see* **Note 11**).
8. (Optional) Wash 1× in sterile PBS, centrifuge at 5000 × *g* for 15 min.
9. Add 10 mL sterile PBS and gently resuspend pellet (*see* **Note 12**).
10. Measure $OD_{600}$ of the final culture and dilute to desired concentration if necessary to achieve the target presented dose.
11. Serially dilute and plate on HIA with Congo Red to verify nebulizer starting material concentration.

### 3.2 Aerosol Challenge

1. All procedures involving live animals must be approved by the institutional animal care and use committee. The procedures described herein were performed in accordance with the Office of Laboratory Animal Welfare and Guide for the Care and Use of Laboratory Animals, and were approved by the University of Missouri Animal Care and Use Committee.
2. Weigh animals ahead of time in order to estimate the required nebulizer concentration that will be needed to achieve the target presented dose.
3. Prepare restraint devices ahead of time, including disinfecting/decontaminating, washing and drying. Lubricate O-ring seals on the pushers/plungers as needed.
4. Operate all aerosol generating and exposure equipment in a class II or III biosafety cabinet (*see* **Note 13**).
5. Confirm proper equipment setup for all supply and exhaust airline connections, impingers, nebulizer, peristaltic pump, thermometer, hygrometer, and any other equipment needed for aerosol generation (Fig. 1).
6. If particle size is to be measured, install and operate a Particle Sizing Spectrometer according to manufacturer's instructions (*see* **Note 14**).
7. Place culture in SLAG nebulizer reservoir if configured for recirculation. Otherwise, inoculum will be pumped from a 50 mL conical tube.
8. Place HIB, with or without antifoam, in impingers (*see* **Note 15**).
9. Set supply flow rate depending on animal mass and number of animals (ports) in the study (Table 1) (*see* **Note 16**).
10. Adjust the flow rate on the exhaust mass flow controllers to maintain the desired negative pressure in the tower.

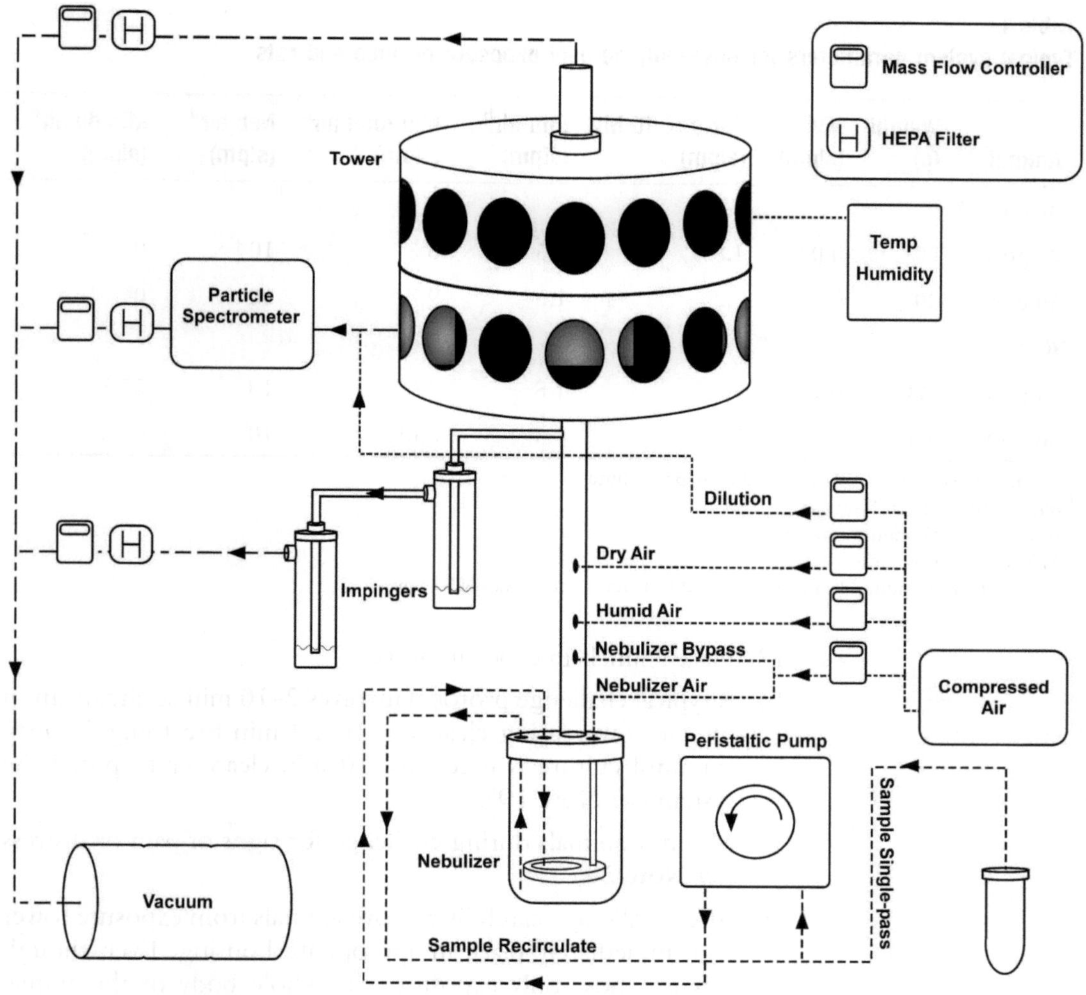


**Fig. 1** Sample setup for nose-only exposure of rodents. Schematic illustrating aerosol generation using compressed air through the SLAG nebulizer. Particle spectrometers, impingers, mass flow controllers, HEPA filters, and temperature/humidity monitors can be attached as indicated. Example shows placement of impingers at the base of the tower in series to sample the aerosol as it enters the exposure tower. Sample single-pass and recirculating options are shown using a peristaltic pump. Biosafety features for working with virulent *Y. pestis* include inline HEPA filters (H) and constant maintenance of negative pressure in the tower (differential pressure gauge for the tower not shown). Dotted lines indicate tubing; arrows indicate direction of air or culture flow

11. Label each restraint device with animal/cage ID and carefully load animals head first, pushing lightly but firmly on backside as they enter. Secure the animal with the pusher/plunger taking care to avoid injury to the tail, and making sure it is facing forward toward the exposure tower port before moving on (*see* **Notes 17** and **18**).
12. Verify all settings for flow rates, negative pressure, and humidity.

**Table 1**
**Typical system parameters for nose-only aerosol exposure of mice and rats**

| Animal | Weight (g) | MV[a] (slpm) | Ports × 10 MV (slpm) | Imp air[b] (slpm) | Min total air[c] (slpm) | Neb air[d] (slpm) | Min dil air[e] (slpm) |
|---|---|---|---|---|---|---|---|
| *Mouse* | | | | | | | |
| 24 port | 20 | 0.02 | 4.8 | 1.5 | 6.3 | 10 | 0 |
| 36 port | 20 | 0.02 | 7.2 | 1.5 | 9.7 | 10 | 0 |
| *Rat* | | | | | | | |
| 24 port | 300 | 0.15 | 36 | 1.5 | 37.5 | 10 | 27.5 |
| 36 port | 300 | 0.15 | 54 | 1.5 | 55.5 | 10 | 45.5 |

[a]*MV* minute volume, *SLPM* standard liters per minute
[b]*Imp air* Impinger airflow rate
[c]*Min total air* Minimum total airflow
[d]*Neb air* Nebulizer airflow rate
[e]*Min dil air* Minimum dilution air required to maintain ten times MV of air

13. Place animals in exposure tower.
14. A typical challenge protocol involves 2–10 min acclimatization on the tower using clean air, 10–20 min breathing *Y. pestis* (or until culture is used up), 10 min clean air to purge the system (*see* **Note 19**).
15. Monitor animals during challenge for signs of pain or distress (*see* **Note 20**).
16. After challenge, carefully remove animals from exposure tower and immediately place in appropriate housing. Even though this is a nose-only exposure, the whole body of the animal should be treated as if potentially contaminated with *Y. pestis* (*see* **Note 21**).
17. Thoroughly flush all tubing containing inoculum with 10% bleach followed by water.
18. Surface-decontaminate the restraint devices after removing the animal by spraying with 70% ethanol. Remove fecal material, if any.
19. Restraint devices are first decontaminated by soaking in 10% bleach (0.5% sodium hypochlorite) for 20 min, and then thoroughly rinsed with water. After decontamination the restraint devices are washed in a dishwasher and stored for future use.
20. Serially dilute and plate impinger samples.
21. Decontaminate exposure system (including biosafety cabinet) using vaporous hydrogen peroxide generator and 30% $H_2O_2$ or similar approved decontamination procedure.
22. Monitor animals for development of disease, according to approved investigator protocol (*see* **Note 22**).

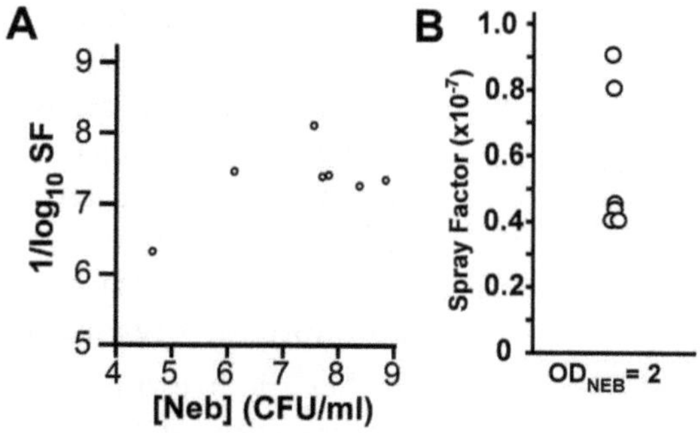


**Fig. 2** Reproducible generation of highly infectious *Y. pestis* C092. (**a**) Calculated aerosol spray factors for high, mid, and low challenge doses showing variation dependent on culture concentration (reproduced from [5]). (**b**) Calculated aerosol spray factors using the standardized protocol for preparation and delivery; nebulizer concentration was approximately $2 \times 10^9$ CFU/mL ($OD_{600} = 2$) for these experiments

### 3.3 Calculation of Presented Dose

1. Impinger and nebulizer concentrations are calculated from serial dilution and plating as indicated above. Aerosol concentration is calculated as the total number of CFU in the impinger media divided by the total volume of air drawn through the impinger.
2. For rodent challenges, the presented dose is calculated as: (aerosol concentration) × (total volume inhaled). Total volume inhaled is the minute volume (MV) calculated using Guyton's formula [7] multiplied by the time of exposure.
3. Spray factor is calculated as aerosol concentration/nebulizer concentration. For a given series of experimental conditions the spray factor should remain consistent between experiments to allow for accurate prediction of presented dose (Fig. 2) (*see* **Note 23**).

## 4 Notes

1. We use this media for short- and long-term storage at −80 °C. Others based on DMSO or glycerol may also work.
2. The sparging liquid aerosol generator (SLAG) was developed as a gentler method of generating infectious aerosols. With the SLAG, aerosols are generated by delivering culture material using a peristaltic pump onto a porous metal disk (frit), while simultaneously forcing air through the frit from below. This causes bubbles which burst, generating infectious aerosol particles which are then carried by the airstream. Similar spray factors and particle size can be achieved as other common aerosol generators such as the Collison jet nebulizer. The

Collison jet nebulizer, which originated for conducting studies on inhaled toxic compounds, can also be used for generation of infectious *Y. pestis* aerosol. The Collison nebulizer works by forcing air through jets which then draw up culture that mixes with air and is impacted against the sides of the reservoir. The rapid acceleration and deceleration of this method may cause loss of microbial viability, especially concerning for the non-spore-forming organisms such as *Yersinia pestis* [8].

3. Smaller frit sizes may bias particle size to give higher aerosol concentration at lower flow rates; however, for higher total airflow, larger pore sizes should be used.
4. Use impingers and other sampling devices according to manufacturer's instructions. Air drawn through impingers should be filtered before passing through vacuum pump. Recommended sampling flow rates vary widely between devices. We typically use our Teflon impingers at 1.5 slpm with approximately 90% collection efficiency as measured using impingers in series (Fig. 1).
5. Many antifoam products are commercially available. We have used Anti-foam B at 0.1% v/v. Foaming can also be reduced by using PBS or other nonfoaming media in the impingers or reducing impinger flow rate. Various media and flow rates should be tested prior to conducting the animal challenge to maximize the efficiency of sampling. This can be done by placing impingers in series. The impinger that is proximal to the exposure tower should capture at least 90% of the challenge aerosol.
6. Animals should fit snugly but not too tight; should not be able to turn away from port.
7. Mass flow controllers facilitate accurate control of airflows. Our system has mass flow controllers for main tower vacuum, aerodynamic particle sizer (APS) sampling and dilution, humidified and nonhumidified dilution air, and nebulizer and impinger flow rates (depicted in Fig. 1).
8. Variations in culture growth conditions and inoculum preparation can affect spray factor and therefore presented dose. To minimize variability in spray factor and improve dosing accuracy, it is important to standardize protocols for culturing and harvesting *Y. pestis*. We strongly recommend using well-characterized working stocks for inoculation of challenge material. This will improve experiment-to-experiment consistency as well as reduce lead time for challenges.
9. Empirically determine CFU/mL in 1 $OD_{600}$ unit to estimate concentration and volume required to inoculate cultures. We find a 1:500 dilution from frozen working stocks (frozen at a

final $OD_{600}$ of 0.2) yields log phase growth in about 24 h when grown as indicated.

10. Growth at 37 °C in 2.5 mM $CaCl_2$ is used to ensure that bacteria are primed for the upregulation of important temperature-regulated virulence factors, including but not limited to CaF1 capsule, type III secretion system (T3SS), and tetraacylated lipid A. The addition of $CaCl_2$ to the media prevents the so-called low calcium response whereby growth restriction is caused by high level expression of the T3SS. The amount of time needed for growth should be standardized and is based on the volume grown, desired target dose, and concentration of the working stock.
11. Increased time and/or rotor speed may increase yield; however, in our experience spray factor is negatively impacted by increased sedimentation.
12. For most nebulizers, 10 mL is sufficient volume for aerosolization and animal challenge. Actual volume should be determined based on individual needs including peristaltic pump flow rate and time of exposure. Decreased flow rates or increased volume of culture can be used to extend exposure time if necessary.
13. Aerosolization of wild type *Y. pestis* is associated with significant risk to human health and must be performed in a select agent authorized BSL3 facility in compliance with all applicable regulations. Consult with a biosafety professional to identify risks and adopt appropriate PPE and biosafety practices. Our exposure system is housed in a class III biosafety cabinet (BSC) in a select agent-registered BSL3 facility. In addition, the exposure system is under negative pressure and the class III BSC is exhausted through a dedicated, double HEPA-filtered exhaust system. The class III BSC is connected via a wall-mounted airlock to a class II BSC in the ABSL3 animal holding room. Animals are passed to and from the class III BSC without leaving containment.
14. Most nebulizers reproducibly generate particles of a given size range, so it may not be necessary to measure particle size in every experiment. If particle size information is desired, there are many instruments available for measuring particle size and concentration. Particle size measurement can easily be configured into aerosol exposure systems and used as described in Subheading 3.2, **step 4** (example shown in Fig. 1).
15. The media in the impingers may foam. Foaming is more severe at higher flow rates and in rich media. To minimize foaming, use reduced flow rates and minimal media or add antifoam.
16. As a general rule we deliver ten times the minute volume requirement of animals. Total flow should be calculated based

on the minute volume (MV) required by each animal times the total number of ports (both used and unused) on the exposure tower times the number of minute volumes desired (*see* Table 1 for examples).

17. Loading animals in restraint devices should proceed simultaneously with culture preparation such that the time the animals spend under restraint in minimized. Animals should be restrained no more than 60 min. For each challenge, we typically have two people preparing/operating the aerosol system and two people loading and unloading animals.
18. While loading, it is relatively common for the animal to turn around in the restraint device before being secured. When this happens, we have found it is easier to leave the animal in the device, loosely positioned, until it self-corrects its position. Upon self-correcting, the restraint device can be closed to secure the animal face-forward. It is generally not advisable to remove the animal from the restraint device for reloading, as it may become less cooperative.
19. For small increases in presented dose exposure time can be increased (if approved by the institutional animal care and use committee) provided there is enough culture material to support the increased duration. However, if a lower dose is desired it is recommended to decrease the concentration of the inoculum rather that shorten the exposure to less than 10 min.
20. It is very important that the animals fit properly in the restraint chamber. This facilitates loading, reduces animal stress, and restricts movement which reduces the potential for animal injury and may improve lower lung deposition of challenge material. If the animal turns around during challenge, remove the restraint device from the tower and reposition the animal as soon as possible. Return the animal to the tower and continue the exposure. Note the amount of time the animal was off the tower to account for the reduced exposure time during calculation of presented dose.
21. Animals typically tolerate the procedure very well without acclimatization to the restraint devices. Animals may be briefly disoriented when then are removed from the restraint devices but typically return to normal quickly.
22. Overt signs of disease typically develop in rodents 2–4 days following challenge.
23. For the SLAG nebulizer and the culture conditions outlined above, we typically observe a spray factor of 0.5–1 $\times 10^{-7}$, given the culture conditions outlined above and a starting nebulizer concentration of approximately $2 \times 10^{9}$ CFU/mL.

## Acknowledgments

We are grateful for excellent BSL3 assistance from Jacob Stockton. This work was supported by NIH/NIAID Public Health Service Award #R21AI133450 (DMA and PEA).

### References

1. Turner P, Brabb T, Pekow C, Vasbinder M (2011) Administration of substances to laboratory animals: Routes of administration and factors to consider. J Am Assoc Lab Anim Sci 50 (5):600–613
2. Hofstetter C, Flondor M, Boost K, Koehler P, Bosmann M, Pfeilschifter J, Zwissler B, Muhl H (2005) A brief exposure to isoflurane (50 s) significantly impacts on plasma cytokine levels in endotoxemic rats. Int Immunopharmacol 5:1519–1522
3. Christopher A, Ochoa S, Krushansky E, Francis R, Tian X, Zahid M, Munoz R, Lo C (2014) The effects of temperature and anesthetic agents on ciliary function in murine respiratory epithelia. Front Pediatr 2:111
4. Gargiulo S, Greco A, Gramanzini M, Esposito S, Affuso A, Brunetti A, Vesce G (2012) Mice anesthesia, analgesia, and care, Part 1: anesthetic considerations in preclinical research. ILAR J 53 (1):E55–E69
5. Gater S, Peters K, Kocsis A, Dhariwala M, Anderson D, Anderson P (2012) Host stress and immune responses during aerosol challenge of Brown Norway rats with *Yersinia pestis*. Front Cell Infect Microbiol 2:147
6. Surgalla M, Beesley E (1969) Congo red-agar plating medium for detecting pigmentation in *Pasteurella pestis*. Appl Microbiol 18 (5):834–837
7. Guyton A (1947) Measurement of the respiratory volumes of laboratory animals. Am J Phys 150:70–77
8. Mainelis G, Berry D, An H, Yao M, DeVoe K, Fennell D, Jaeger R (2005) Design and performance of a single-pass bubbling bioaerosol generator. Atmos Environ 39(19):3521–3533

# Chapter 4

# Mouse Models of Yersiniosis

**Diana Hooker-Romero, Leah Schwiesow, Yahan Wei, and Victoria Auerbuch**


## Abstract

Yersiniosis is common foodborne gastrointestinal disease caused by the enteric pathogens *Yersinia enterocolitica* and *Yersinia pseudotuberculosis*. The mouse model of oral infection serves as a useful tool to study enteropathogenic *Yersinia* infection in mammals. The following protocol describes two distinct oral infection methods: the commonly used oral gavage method in which the bacterial inoculum is instilled directly into the mouse stomach using a feeding needle, and an alternative method in which mice are fed bread soaked with *Yersinia* culture.


**Key words** Yersiniosis, *Yersinia pseudotuberculosis*, *Yersinia enterocolitica*, Oral gavage, Gastroenteritis, Foodborne infection model, Bread-feeding model

## 1 Introduction

Yersiniosis is a foodborne, self-limiting gastrointestinal disease caused by the gram-negative bacterial pathogens *Yersinia enterocolitica* and *Yersinia pseudotuberculosis*, which are usually transmitted via the fecal–oral route [1]. Although both *Y. pseudotuberculosis* and *Y. enterocolitica* cause disease, most cases in humans are associated with *Y. enterocolitica* [1]. In otherwise healthy individuals, yersiniosis is characterized by mild, self-limiting diarrhea, enteritis, ileocolitis, and mesenteric lymphadenitis, and can affect the appendix tissue and mimic appendicitis [1]. In immunocompromised patients, or in patients with iron overload [2], *Yersinia* can disseminate to the spleen and liver. The *Yersinia* mouse model of oral infection is the most widely used animal model to study enteric *Yersinia* infections, as it reflects the natural route of infection where the pathogen is exposed to the mucosal immune environment of the small intestine before disseminating into deeper tissues [3, 4]. Specifically, the mouse model of oral infection mimics human disseminated infection in that the spleen and liver become colonized.

Viveka Vadyvaloo and Matthew B. Lawrenz (eds.), *Pathogenic Yersinia: Methods and Protocols*, Methods in Molecular Biology, vol. 2010, https://doi.org/10.1007/978-1-4939-9541-7_4, © Springer Science+Business Media, LLC, part of Springer Nature 2019

The oral infection model was first established by Phil Carter and Frank Collins in 1974 when they demonstrated that after oral gavage, mice displayed similar symptoms to those observed in humans, such as formation of abscesses in Peyer's patches, mesenteric lymph nodes, spleen, and liver [5, 6]. These abscesses can resolve in animals that survive the infection [5–8]. After ingestion, *Yersinia* adheres to the apical surface of microfold (M) cells within the small intestinal lining, where the bacteria cross the epithelial barrier to the underlying Peyer's patches [9, 10]. Within the Peyer's patches, *Yersinia* replicates extracellularly and disseminate to distal lymphoid tissues such as mesenteric lymph nodes [1]. However, *Yersinia* can also replicate in the intestinal lumen and a Peyer's patch independent route of dissemination from the gut lumen to the spleen and liver has been described for *Y. pseudotuberculosis*, although the mechanism remains unclear [11]. This Peyer's patch independent route enables successful colonization of the spleen and liver, where monoclonal microabscesses are formed [7, 11].

Numerous studies using the mouse oral infection model have identified a number of virulence factors needed for infection and survival in the host [8, 10, 12–20]. In addition, a recent study showed that virulence factor expression can vary within an abscess, depending on proximity to immune cells and other parameters [8]. Other studies have provided a better understanding of how the host immune system recognizes and reacts to *Yersinia* [17, 19, 21–27]. The dynamic interplay between a variety of host cells and *Yersinia* capable of sensing environmental cues to control virulence gene expression underscores the importance of studying the host–pathogen interaction in the context of a natural infection model. Studies like the one by Nuss et al., which analyzed the interaction between *Y. pseudotuberculosis* and the murine host by performing a dual RNAseq technique on infected mouse Peyer's patches [26], will help fill in the gaps of our understanding.

The following protocols describe two different methods of orally inoculating mice with enteropathogenic *Yersinia*. The first method, oral gavage, which introduces the bacterial inoculum directly into the stomach, is the most commonly used procedure. The second method, bread feeding, is a new protocol recently developed for *Listeria monocytogenes* infection by Ghanem et al. (2013), in which bacterial culture is resuspended in melted butter and soaked into a small piece of white bread [28]. Our lab has adapted this method to use with *Y. pseudotuberculosis* infection of mice because it resembles a more natural route of oral infection and prevents possible physical trauma to the lining of the esophagus, which could cause direct entrance of the inoculum into the bloodstream [28]. This method also limits stress on the animal subjects during delivery of the infectious dose. However, one limitation is that, depending on the mouse genotype and susceptibility to

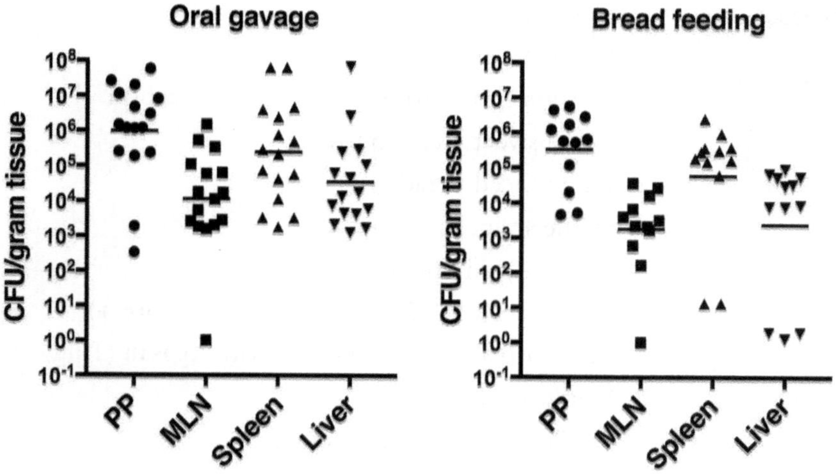


**Fig. 1** Bacterial load in the different organs is comparable after oral infection of *Yersinia pseudotuberculosis* by oral gavage or bread feeding. Mice were infected with $2 \times 10^8$ CFU of WT *Y. pseudotuberculosis* IP2666 and after 5 days the mesenteric lymph nodes (MLN), Peyer's patches (PP), livers, and spleens were harvested, homogenized, and plated for CFU determination. Oral gavage data points were previously published in Miller et al. [20, 21]

*Yersinia* infection, it can be challenging to deliver a large enough inoculum. Therefore, we have performed this method successfully in 129S1/SvImJ mice (*see* Fig. 1) but not in C57Bl/6 mice, which require a higher dose [4].

The oral mouse models of infection described here can be used to determine virulence kinetics of different *Y. pseudotuberculosis* or *Y. enterocolitica* strains or mutants. Dissemination can be analyzed by plating homogenates of mouse organs and counting colony-forming units (CFU) or by sectioning mouse organ tissue and analyzing microabscess formation by immunohistochemistry. The latter approach is usually done after infecting mice with fluorescently labeled bacteria [7, 8, 19].

## 2 Materials

### 2.1 Oral Gavage

1. Sterile 1 ml syringes and gauge 10, stainless-steel 2-in. feeding needles with a smooth, 3 mm ball on the end.
2. Luria broth (LB) medium.
3. Sterile 1× Phosphate-buffered saline (1× PBS).
4. *Yersinia* strains to be used during infection (*see* **Note 1**).
5. 8–10 week old 129S1/Sjv mice (*see* **Notes 2 and 3**).
6. LB agar plates supplemented with irgasan (1 μg/ml).
7. Sterile 1× PBS.
8. Sterile spreading glass beads.
9. 1.5 ml Eppendorf tubes.

### 2.2 Oral Infection via Ingestion of Contaminated Bread

1. Sterile 1× PBS.
2. LB medium.
3. *Yersinia* strains to be used during infection.
4. Salted sweet cream butter.
5. White sliced bread.
6. Sterile scalpel.
7. Sterile 1.5 ml microcentrifuge tubes.
8. 8–10 week old 129S1/Sjv mice (*see* **Note 4**).
9. LB agar plates supplemented with irgasan (1 μg/ml).
10. Sterile 1× PBS.
11. Sterile spreading glass beads.
12. 1.5 ml Eppendorf tubes.

### 2.3 Dissemination Analysis

1. 15 ml conical tubes.
2. LB agar plates supplemented with irgasan (1 μg/ml). Approximately six plates per organ per mouse.
3. Cotton balls.
4. Sterile 1× PBS.
5. Isoflurane.
6. Sterile spreading glass beads.
7. Sterile 1.5 ml Eppendorf tubes.
8. Sterile scalpel, forceps, and scissors for mice tissue dissection.
9. Sterile 5 ml culture tubes or 15 ml conical tubes to collect mice tissues and perform homogenization.
10. Scale.
11. Plastic Tupperware with cover.
12. Omni Tissue Homogenizer with stainless steel probe (125 W, speed 5000–35,000 rpm).
13. 70% (v/v) ethanol.
14. 100% ethanol.
15. 50 ml conical tubes.
16. Fresh 4% paraformaldehyde in 1× PBS.
17. 30% sucrose in 1× PBS.
18. Leica cryostat.
19. OCT compound-freezing media.
20. Positively charged slides.
21. ProLong Gold Antifade Mountant.
22. Clear nail polish.

23. Coverslips.
24. 2-methylbutane
25. Dry ice.

## 3 Methods

### 3.1 Oral Gavage

1. 15–18 h prior to infection, fast mice by removing chow. Continue providing water ad libitum (*see* **Note 5**).
2. 15–18 h prior to infection, inoculate 2 ml of LB media with *Yersinia* strain. Grow at 26 °C with agitation.
3. Take OD of the overnight culture and calculate amount of culture needed to inoculate $2 \times 10^8$ bacteria per mouse (*see* **Note 6**). Transfer culture to Eppendorf tube and spin down for 5 min at $835 \times g$. Decant supernatant and resuspend bacterial culture in 1× PBS to approximately $2 \times 10^9$ bacteria/ml. Each mouse will get 200 μl aliquots of $2 \times 10^8$ bacteria per inoculum (*see* **Note 7**).
4. Uptake the bacterial suspension into the syringe and position the feeding needle securely.
5. Restrain the mouse securely by the scruff and hold mouse in an upright position, preventing the mouse from moving its head (*see* **Note 8**).
6. Locate the mouse's last rib; this is the position of the stomach. With the ball tip at the last rib, place the needle alongside the mouse and look at where the shaft meets the mouse's incisors; this is the point at which you would stop advancing the needle.
7. Introduce the feeding needle into the mouse mouth slowly and guide the ball tip along the roof of the mouth and toward the back of the throat, incline the syringe so that it is parallel with the mouse body and gently guide the needle down the esophagus. Never force the needle. Remove the needle and try again if you get resistance. Once the needle tip is in the stomach, inject 200 μl slowly. Control mice can be inoculated with sterile 1× PBS.
8. Remove the needle straight up slowly.
9. Return mouse to cage and monitor for any signs of respiratory distress or injury, which would indicate unintended injury during gavage.
10. Provide food and water *ad libitum*.
11. Verify inoculum by plating tenfold serial dilutions onto LB plates with irgasan 1 μg/ml, as it selects for *Yersinia* species. For the dilutions, add 900 μl of 1× PBS into 1.5 ml Eppendorf tubes. Take 100 μl of inoculum and pipet into the first

Eppendorf tube, vortexing vigorously for 15 s, for a 1:10 dilution. Take 100 μl of the 1:10 dilution and add it to the next tube containing 900 μl sterile 1× PBS to make the 1:100 dilution. Repeat these steps to make five dilutions for each organ.

12. Add 100 μl of each dilution to a previously labeled LB plate supplemented with irgasan, add spreading beads and shake plate for 30 s to spread the bacteria. Remove the beads.
13. Incubate plates at 26 °C for 48 h, then count colonies.

### 3.2 Oral Infection via Ingestion of Contaminated Bread

1. 15–18 h prior to infection, fast mice by removing chow. Continue providing water ad libitum.
2. 15–18 h prior to infection, inoculate 2 ml of LB media with *Yersinia* strain. Grow at 26 °C with agitation.
3. Remove crust from bread and cut into small 3–4 mm cubes with the sterile scalpel blade (*see* **Note 9**) (*see* Fig. 2).
4. Cut butter into 0.5–1 cm cubes and place into 1.5 ml microcentrifuge tube (*see* **Note 9**).

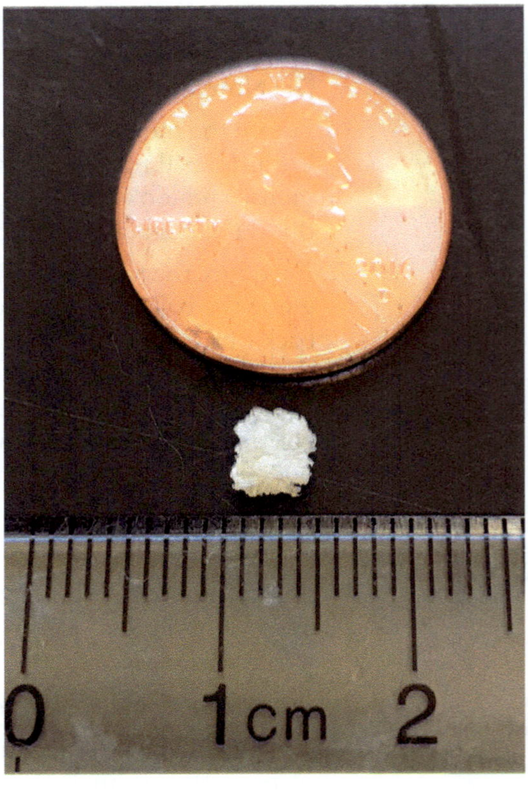


**Fig. 2** Measurement of the bread size for mouse infection. After cutting the crusts, the bread is sliced with a sterile scalpel. Pieces should be cubes of about 3–4 mm$^3$

5. Incubate microcentrifuge tube with 2 ml aliquot of 1× PBS in water bath at 55 °C.
6. Incubate microcentrifuge tube with previously cut butter in water bath at 55 °C.
7. Once the overnight culture is ready, take OD and calculate number of bacteria needed for an infectious dose of $2 \times 10^8$ per mouse. For example, to inoculate 10 mice, transfer $2 \times 10^9$ cells to an Eppendorf tube.
8. Spin down culture at 835 × *g* for 5 min, then aspirate supernatant.
9. Each inoculum needs to be resuspended in 2 μl of previously incubated at 55 °C 1× PBS and 3 μl of melted butter. For 10 inoculums, resuspend the $2 \times 10^9$ cells in 20 μl of warm 1× PBS and 30 μl of melted butter. Mix thoroughly and rapidly pipet the 5 μl onto a bread cube (*see* **Note 10**). Store tubes with inoculated bread on ice until ready to do infection.
10. Place each mouse in an empty cage and provide the bread piece. Wait until the animal has eaten the entire piece of bread. This usually takes a few minutes (*see* **Note 11**).
11. After infection, return mouse to original cage.
12. Provide food and water *ad libitum*.
13. Verify inoculum by placing an inoculated bread piece in 1 ml of warm 1× PBS and vortexing vigorously for 30 s.
14. Serially dilute bread homogenate by a tenfold factor by adding 100 μl of it into 900 μl of 1× PBS. Continue serially diluting to $10^{-7}$.
15. Plate the last four dilutions onto LB plates supplemented with irgasan by pipetting 100 μl and spreading with sterile glass beads.

### 3.3 Methods for Studying Bacterial Dissemination

After inoculation, check mice daily for signs of distress. Inspect daily for any inactivity, lack of responsiveness to stimulation, reduced eating or drinking, hunched posture, or ruffled fur, which can be indicative of severe illness. If mice display any of those signs, euthanize immediately to avoid suffering. Tissues can be harvested between 1 and 5 days postinoculation.

#### 3.3.1 Bacterial Enumeration in Infected Tissues

1. Label and weigh one 15 ml conical tube for each tissue that will be harvested.
2. Euthanize mice by an American Veterinary Medical Association-approved method.
3. Pin mouse to dissecting board and spray ventral side with 70% ethanol.
4. Make an incision along the abdomen with a sterile scissors.

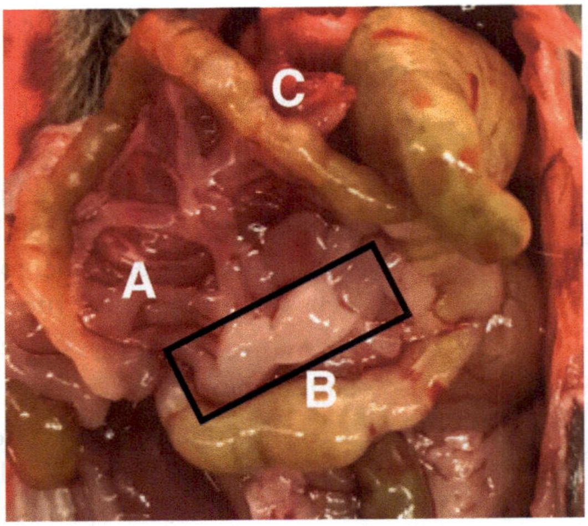


**Fig. 3** Location of the inflamed mesenteric lymph nodes and Peyer's patches. Once the intestines are exposed, lift the intestinal coils on the mouse's right side and shift over to the left to find the mesentery (**a**). Find the mesenteric lymph nodes nestled in the mesentery (**b**). They will look like small white oval plump structures. The Peyer's patches are located along the intestine walls (**c**). They look like pearly white raised dots on the surface of the small intestine

5. Place intestines on one side, without damaging the mesentery, so that the mesenteric lymph nodes are exposed (*see* **Note 12**) (*see* Fig. 3).
6. Using aseptic technique, remove the mesenteric lymph nodes and place them on ice in a previously weighed conical tube.
7. Remove the Peyer's patches along the small intestine (*see* Fig. 3) by using a scissors to cut each Peyer's patch away from the small intestinal tissue, taking care not to include luminal contents. Place the Peyer's patch in a previously weighed conical tube. Incubate on ice. Peyer's patches belonging to the same mouse are placed into the same tube.
8. Remove spleen and liver and place on ice in individual previously weighed conical tubes (*see* **Note 13**).
9. Weigh the tubes containing the organs.
10. Add 1 ml of sterile 1× PBS to the tubes holding the livers and 500 μl to the tubes holding the remaining organs (*see* **Note 14**).
11. Homogenize the organs for 30 s with tissue homogenizer at 20,000 rpm. Smaller organs such as Peyer's patches and mesenteric lymph nodes are homogenized first, then the spleen is homogenized, followed by the liver. Homogenization should be done in a laminar air flow hood since bacteria can be aerosolized. Wash homogenizer after each set of organs by running the homogenizer at 20,000 rpm for 15 s in a 50 ml

conical tube filled with 1× PBS, followed by a run in 100% ethanol and then 1× PBS again. A fresh set of washing tubes is used for each mouse.

12. Serially dilute homogenate by a tenfold factor by adding 100 μl of it into 900 μl of 1× PBS. Continue serial dilution from $10^{-1}$ to $10^{-4}$.
13. Use undiluted homogenate as well as the $10^{-1}$ to $10^{-4}$ dilutions for Peyer's patches, mesenteric lymph nodes, liver, and spleen. Plate onto LB plates supplemented with irgasan by pipetting 100 μl and spreading it with sterile glass beads.
14. Incubate plates at 26 °C for 48 h.
15. Count colony forming units (CFU) and calculate CFU per gram of tissue using the following formula (*see* Fig. 1):

$$\mathrm{CFU}/_{\mathrm{gr}}\ \mathrm{Organ} = \left(\frac{\mathrm{CFU}}{\mathrm{Vol.plated}}\right)\left(\frac{\mathrm{Total\ vol.PBS}}{\mathrm{Weight\ of\ organ}}\right)(\mathrm{Dilution\ factor})$$

*3.3.2 Immunofluorescence Analysis of Infected Tissues*

If mice have been infected with fluorescent *Yersinia* strains, bacterial dissemination can also be examined by microscopy.

1. Harvest tissues as described in Subheading 3.3.1, **steps 1–4**.
2. Place organs in 15 ml conical tubes containing 1 ml 1× PBS. Swirl the tube slowly for 15 s, decant PBS, and repeat wash with fresh PBS two more times.
3. Fix the organs by placing them in a 15 ml conical tube and adding 1 ml of fresh 4% PFA in PBS. Incubate for 3 h.
4. Wash tissue by transferring each organ to a new 15 ml conical tube with 1 ml 1× PBS and incubating for 5 min at room temperature. Repeat once more.
5. Incubate in 2 ml of 30% sucrose overnight at 4 °C.
6. To freeze-embed the tissue, add about 2 ml of tissue freezing media to a cryogenesis cuvette, then add the organ and completely cover with more freezing media. Make sure to add media slowly to avoid any bubbles.
7. Submerge the cuvette in a flask filled with ~200 ml of 2-methylbutane over dry ice (*see* **Note 15**).
8. Cut 10 μm sections with a cryostat microtome and affix to slides. Store slides in dark slide box at 4 °C until ready to stain.
9. Stain tissues with Hoechst at a 1:10,000 dilution in 1× PBS. Incubate for 30 min in the dark at room temperature.
10. Wash two times by tilting the slide to 45° and slowly pipetting 1× PBS to the top of the slide and letting it drip over the tissue slice.

11. Affix coverslips with an antifade mountant such as ProLong Gold and seal the sides of the coverslip with clear nail polish.
12. Let the ProLong Gold cure and nail polish dry overnight by storing in the dark at room temperature.
13. Examine by fluorescence microscopy.

## 4 Notes

1. *Y. pseudotuberculosis* and *Y. enterocolitica* are considered potentially hazardous. Biosafety Level 2 practices and facilities are required when working with these pathogens.
2. All experimental procedures using live animals must be approved by the relevant Institutional Animal Care and Use Committee (IACUC).
3. Yersiniosis oral model studies usually use mice between 7 and 12 weeks of age, as infant mice do not develop systemic infection [29]. The three main inbred mouse strains used for *Yersinia* oral infection are C57BL/6, BALB/c, and 129S1/Sjv. A study comparing the three mouse strains showed no difference in dissemination patterns and tissue pathology; however, mesenteric lymph nodes of C5BL/6 mice displayed a lower bacterial load, compared to the other two strains. Additionally, this study also reported high variability in spleen colonization in C57BL/6 and BALB/c mice [4]. $LD_{50}$ analysis showed that BALB/c mice are more susceptible and that C57BL/6 mice are more resistant.
4. We have performed this method with 8–12-week-old 129S1 mice only and have not tested the optimal inoculum for other mouse inbred strains. As the infectious dose used to infect C57Bl/6 is higher due to higher resistance, the pieces of bread fed to these animals would need to be larger to soak up the larger *Yersinia* pellet resuspension. This can be challenging given the small size of the mouse stomach. Additionally, a bigger bacterial pellet might also be hard to resuspend in such a small amount of butter, making the suspension very sticky and hard to get off of the tube.
5. Sodium chloride or sodium bicarbonate can be supplemented before inoculation. Pretreatment of mice with sodium bicarbonate can reduce gastric acid and thus resistance to gastric infection [30]. However, another study giving different concentrations of sodium chloride to mice before orally inoculating them with *Y. enterocolitica* suggested that neutralization of gastric acidity had no effect on bacterial virulence [31].
6. Quantity of cells corresponding to a specific OD can be calculated by growing an overnight culture of the strains to be

inoculated, taking the OD with a spectrophotometer and then performing a tenfold serial dilution, from which 100 μl are plated on LB plates. CFU can be counted after 48 h incubation at 26 °C.

7. Different studies using *Yersinia pseudotuberculosis* strains IP2666, YPIII, and IP32953 have shown that the ideal inoculum is between $1 \times 10^8$ to $8 \times 10^8$ CFU for C57BL/6 and 129S1/Sjv mice, and about $2 \times 10^7$ for BALB/c mice. However, dose depends on bacterial serotype, as some serotypes can be more virulent than others.

8. Mice can be lightly anesthetized using a vaporizer to deliver a small dose of isoflurane before performing oral gavage. This can minimize stress on the mice and prevent mice from moving during gavage. After gavage, mice are monitored for several minutes after they regain consciousness to ensure they do not show any signs of distress or discomfort. Isoflurane usage should be performed in a ducted biosafety cabinet or in the presence of an isoflurane scavenger system to avoid accidental inhalation.

9. Store individual bread pieces in a microcentrifuge tube or small petri dish at −20 °C. This can be done up to a day before inoculation of the bread. Butter can also be stored ahead of time in an Eppendorf tube at −20 °C.

10. To avoid fast cooling of the butter, bacterial resuspension in warm butter and PBS can be done in batches. Avoid doing more than ten pieces at the time or the butter will solidify. To keep the butter melted in between bread inoculations, rapidly dip bacterial suspension tube in a 55 °C water bath for a couple seconds and pipette several times to mix bacterial suspension. Make sure this is done rapidly so as not affect bacterial viability.

11. If mouse does not eat the piece of bread within 15 min, mouse can be left undisturbed in the cage for up to 2 h.

12. Mesenteric lymph nodes are usually removed first before disrupting the mesentery. Once the intestines are exposed, lift the intestinal coils on the mouse's right side and shift over to the left to find the mesentery. Find the mesenteric lymph nodes nestled in the mesentery. They will look like small white oval plump structures (*see* Fig. 3).

13. It is also possible to analyze *Yersinia* within the small intestine and feces. When plating the small intestine, squeeze out the contents with forceps and plate that separately from the small intestinal tissue. Cecal content and fecal pellets can also be homogenized and analyzed for bacterial content.

14. PBS can be added prior to weighing the tubes. NP-40 detergent can also be added to PBS to lyse open the mouse cells. No

difference between the use of PBS and PBS with NP-40 has been found in terms of CFU [27].

15. 2-methylbutane should be previously incubated at −70 °C the night before tissue embedding.

## Acknowledgments

We thank Susannah McKay and Daniel Portnoy for technical advice in establishing the bread feeding model in *Yersinia*. D.H.-R. is supported by a Ford Foundation fellowship. The mouse infection data shown are taken from the work supported by the National Institute of Allergy and Infectious Diseases of the National Institutes of Health under Award Numbers R21AI099747 and R01AI119082 (to V.A.).

### References

1. Galindo C, Rosenzweig J, Kirtley M et al (2011) Pathogenesis of *Y. enterocolitica* and *Y. pseudotuberculosis* in human yersiniosis. J Pathogens 2011:182051
2. Weinberg E (2000) Microbial pathogens with impaired ability to acquire host iron. Biometals 13:85–89
3. Lawrenz M (2010) Model systems to study plague pathogenesis and develop new therapeutics. Front Microbiol 1:119
4. Handley S, Dube P, Revell P et al (2004) Characterization of oral *Yersinia enterocolitica* infection in three different strains of inbred mice. Infect Immun 72:1645–1656
5. Carter P (1975) Animal model of human disease. *Yersinia* enteritis. Animal model: oral *Yersinia enterocolitica* infection of mice. Am J Pathol 81:703–706
6. Carter P, Collins F (1974) Experimental *Yersinia enterocolitica* infection in mice: kinetics of growth. Infect Immun 9:851–857
7. Trülzsch K, Oellerich M, Heesemann J (2007) Invasion and dissemination of *Yersinia enterocolitica* in the mouse infection model. Adv Exp Med Biol 603:279–285
8. Davis K, Mohammadi S, Isberg R (2015) Community behavior and spatial regulation within a bacterial microcolony in deep tissue sites serves to protect against host attack. Cell Host Microbe 17:21–31
9. Grützkau A, Hanski C, Hahn H et al (1990) Involvement of M cells in the bacterial invasion of Peyer's patches: a common mechanism shared by *Yersinia enterocolitica* and other enteroinvasive bacteria. Gut 31:1011–1015
10. Marra A, Isberg R (1997) Invasin-dependent and invasin-independent pathways for translocation of *Yersinia pseudotuberculosis* across the Peyer's patch intestinal epithelium. Infect Immun 65:3412–3421
11. Barnes P, Bergman M, Mecsas J et al (2006) *Yersinia pseudotuberculosis* disseminates directly from a replicating bacterial pool in the intestine. J Exp Med 203:1591–1601
12. Handley S, Newberry R, Miller V (2005) *Yersinia enterocolitica* invasin-dependent and invasin-independent mechanisms of systemic dissemination. Infect Immun 73:8453–8455
13. Mecsas J, Bilis I, Falkow S (2001) Identification of attenuated *Yersinia pseudotuberculosis* strains and characterization of an orogastric infection in BALB/c mice on day 5 postinfection by signature-tagged mutagenesis. Infect Immun 69:2779–2787
14. Clark M, Hirst B, Jepson M (1998) M-cell surface β1 integrin expression and invasin-mediated targeting of *Yersinia pseudotuberculosis* to mouse Peyer's patch M cells. Infect Immun 66:1237–1243
15. Trülzsch K, Sporleder T, Igwel E et al (2004) Contribution of the major secreted yops of *Yersinia enterocolitica* O: 8 to pathogenicity in the mouse infection model. Infect Immun 72:5227–5234
16. Trcek J, Berschl K, Trülzsch K (2010) In vivo analysis of *Yersinia enterocolitica* infection using luxCDABE. FEMS Microbiol Lett 307:201–206
17. Pepe J, Wachtel M, Wagar E et al (1995) Pathogenesis of defined invasion mutants of

*Yersinia enterocolitica* in a BALB/c mouse model of infection. Infect Immun 63:4837–4848

18. Balada-Llasat J, Mecsas J (2006) *Yersinia* has a tropism for B and T cell zones of lymph nodes that is independent of the type III secretion system. PLoS Pathog 2:e86
19. Crimmins G, Mohammadi S, Green E et al (2012) Identification of MrtAB, an ABC transporter specifically required for *Yersinia pseudotuberculosis* to colonize the mesenteric lymph nodes. PLoS Pathog 8:e1002828
20. Miller H, Kwuan L, Schwiesow L et al (2014) IscR is essential for *Yersinia pseudotuberculosis* type III secretion and virulence. PLoS Pathog 10:e1004194
21. Miller H, Schwiesow L, Au-Yeung W et al (2016) Hereditary hemochromatosis predisposes mice to *Yersinia pseudotuberculosis* infection even in the absence of the type III secretion system. Front Cell Infect Microbiol 6:69
22. Autenrieth I, Kempf V, Sprinz T et al (1996) Defense mechanisms in Peyer's patches and mesenteric lymph nodes against *Yersinia enterocolitica* involve integrins and cytokines. Infect Immun 64:1357–1368
23. Autenrieth I, Beer M, Bohn E et al (1994) Immune responses to *Yersinia enterocolitica* in susceptible BALB/c and resistant C57BL/6 mice: an essential role for gamma interferon. Infect Immun 62:2590–2599
24. Westermark L, Fahlgren A, Fällman M (2014) *Yersinia pseudotuberculosis* efficiently escapes polymorphonuclear neutrophils during early infection. Infect Immun 82:1181–1191
25. Costa T, Amer A, Farag S et al (2013) Type III secretion translocon assemblies that attenuate *Yersinia* virulence. Cell Microbiol 15:1088–1110
26. Nuss A, Beckstette M, Pimenova M et al (2017) Tissue dual RNA-seq allows fast discovery of infection-specific functions and riboregulators shaping host–pathogen transcriptomes. Proc Natl Acad Sci 114:E791–E800
27. Auerbuch V, Golenbock D, Isberg R (2009) Innate immune recognition of *Yersinia pseudotuberculosis* type III secretion. PLoS Pathog 5: e1000686
28. Ghanem E, Myers-Morales T, Jones G et al (2013) Oral transmission of *Listeria monocytogenes* in mice via ingestion of contaminated food. J Vis Exp 74:e50381
29. Echeverry A, Schesser K, Adkins B (2007) Murine neonates are highly resistant to *Yersinia enterocolitica* following orogastric exposure. Infect Immun 75:2234–2243
30. Tennant S, Hartland E, Phumoonna T et al (2008) Influence of gastric acid on susceptibility to infection with ingested bacterial pathogens. Infect Immun 76:639–645
31. Singh A, McFeters G (1987) Survival and virulence of copper and chlorine-stressed *Yersinia enterocolitica* in experimentally infected mice. Appl Environ Microbiol 53:1768–1774

# Part II

## Monitoring Bacteria During Infection

# Chapter 5

# Intravital Confocal Microscopy of Dermal Innate Immune Responses to Flea-Transmitted *Yersinia pestis*

**Jeffrey G. Shannon and B. Joseph Hinnebusch**


## Abstract

The technique known as intravital microscopy (IVM), when used in conjunction with transgenic mice expressing fluorescent proteins in various cell populations, is a powerful tool with the potential to provide new insights into host-pathogen interactions in infectious disease pathogenesis in vivo. *Yersinia pestis*, the causative agent of plague, is typically deposited in a host's skin during feeding of an infected flea. IVM has been used to characterize the innate immune response to *Y. pestis* in the skin and identify differences between the responses to needle-inoculated and flea-transmitted bacteria that would have been difficult, if not impossible, to detect by other means. Here we describe techniques used to image the neutrophil response to flea-transmitted *Y. pestis* in the dermis of live mice using conventional confocal microscopy.


**Key words** Intravital microscopy, Confocal microscopy, Skin, *Yersinia pestis*, *Xenopsylla cheopis*, Flea

## 1 Introduction

The disease plague, caused by the Gram-negative bacterium *Yersinia pestis*, can take three distinct forms, bubonic, pneumonic, and primary septicemic, with bubonic plague being the most common presentation in humans. Plague remains endemic in many countries around the world, where it is maintained in flea-rodent enzootic cycles with occasional outbreaks among humans after contact with infected animals or their fleas. Fleas become infected with *Y. pestis* by feeding on a bacteremic rodent and can transmit the bacteria to another host during subsequent feedings. Fleas deposit *Y. pestis* into the skin, where it adapts to conditions in the mammalian host and survives the host's innate immune response. The bacteria eventually disseminate from the skin, via the lymphatics, to the regional draining lymph node. There, the pathogen replicates to high numbers, resulting in a large, swollen, painful lymph node called a bubo. There is a need for better understanding of host-*Y. pestis* interactions in the skin early after infection and how

Viveka Vadyvaloo and Matthew B. Lawrenz (eds.), *Pathogenic Yersinia: Methods and Protocols*, Methods in Molecular Biology, vol. 2010, https://doi.org/10.1007/978-1-4939-9541-7_5, 

transmission of the bacteria by the natural route, the bite of an infected flea, might affect these interactions.

Numerous published studies describe the use of intravital microscopy to image host-pathogen interactions in the skin in a variety of animal models of infectious disease (reviewed in [1]). The bulk of these studies use multiphoton microscopy due to its ability to image tissues at depths of >100 μm. However, the skin of a mouse ear pinna is very thin, approximately 50–60 μm, allowing for imaging of fluorescently labeled cells and bacteria by conventional single-photon confocal microscopy. The main advantage of this approach is that conventional confocal microscopes are much more widely available than the more sophisticated and costly 2-photon systems. Additionally, even at research institutions that possess multiphoton microscopes, this equipment can be in high demand and getting access for the large blocks of time necessary for intravital experiments can be difficult. Thus, the ability to do this type of experiment with a conventional confocal microscope might allow more researchers to explore the use of IVM to study their model of interest.

A large number of transgenic mice have been created, many commercially available, that constitutively express fluorescent proteins in a variety of cell types relevant to the study of infections in the skin, including neutrophils [2], monocytes/macrophages [3, 4], dendritic cells [4, 5], Langerhans cells [5], and gamma-delta T-cells [6]. Lymphocytes can also be isolated, fluorescently labeled ex vivo, and adoptively transferred into animals to allow intravital imaging [7]. Additionally, fluorophore-conjugated antibodies or chemical labels can be injected into mice prior to IVM, permitting imaging of cell populations or tissues for which the use of transgenic mice is not possible or practical [8].

Here we describe the imaging of bacteria transmitted via the natural route of infection, the bite of an infected flea. These techniques permit the qualitative and quantitative assessment of neutrophil recruitment and bacteria-neutrophil interactions within the skin after *Y. pestis* infection [9, 10]. In addition to their utility in the evaluation of the innate cellular response to *Y. pestis*, these techniques may be adapted to the study of a variety of other pathogens transmitted by other arthropod vectors.

## 2 Materials

### 2.1 Reagents and Equipment

1. Brain-heart Infusion (BHI) broth.
2. Carbenicillin (100 mg/ml in 50% ethanol stock solution).
3. Blood agar plates.
4. Phosphate-buffered Saline (PBS), pH 7.2.

5. Sterile distilled water (dd$H_2O$).
6. Coverslips (60 mm × 24 mm).
7. Coverslip forceps.
8. Membrane feeder device (Wade and Georgi [11]).
9. Heated waterbath.
10. Peristaltic pump and 2 lengths (approx. 1 meter each) of plastic tubing.
11. Flea storage capsules [11].
12. Custom plastic clip-on ear feeding device.
13. Ketamine (13.9 mg/ml)/Xylazine (0.56 mg/ml) solution for injection.
14. Sytox Blue fluorescent DNA stain (ThermoFisher, 1:25 dilution of 1 mM stock solution in PBS).
15. Qtracker 655 vascular dye (ThermoFisher).
16. Microscope stage insert able to accommodate a 60 mm × 24 mm coverslip.
17. Medical adhesive tape.
18. Isoflurane.
19. Electric heating pad.
20. Image analysis software such as Imaris (BitPlane).

### 2.2 Instruments

1. An inverted confocal microscope with incubator chamber surrounding the entire microscope stage and lasers/detectors setup for imaging the desired fluorophores (blue [Sytox Blue], green [eGFP], red [mCherry], and far red [Qdot655 or Alex647]). We use a Zeiss LSR880 confocal microscope system.
2. Dissecting microscope with a chill table and external light source.
3. Isoflurane vaporizer anesthesia apparatus with induction chamber and small rodent breathing circuit and nose cone. We use the RC2 rodent circuit controller (VetEquip).

### 2.3 Organisms

1. Bacterial strains: *Yersinia pestis* strains constitutively expressing fluorescent protein suitable for confocal microscopy. We use the attenuated, BSL2 strain KIM6+ (pCD1 virulence plasmid-negative, pigmentation-positive) of *Y. pestis* transformed with the plasmid pMcherry [9].
2. Mice: This type of study requires an animal-use protocol describing the techniques to be used and approved by an institutional review board. The animal studies described here were performed in accordance with a protocol (#2012-040) approved by the Animal Care and Use Committee, Rocky

Mountain Laboratories, NIAID, NIH. Transgenic mice expressing fluorescent protein in the cell population of interest are required. Here we use LysozymeM-eGFP transgenic mice expressing high levels of eGFP in neutrophils, originally described by Faust et al. [2].

Normal, inbred, or outbred mice to serve as a source of both heparinized mouse blood and fresh mouse skins for flea infections.

3. Fleas: The methods described here use the rat flea *Xenopsylla cheopis*. These fleas were reared in a breeding colony maintained at Rocky Mountain Laboratories. Detailed methods for the establishment and maintenance of flea colonies are described elsewhere [12].

## 3 Methods

### 3.1 Infection of Fleas with *Y. pestis* Expressing mCherry

1. Inoculate 4 ml of BHI broth containing 100 μg/ml carbenicillin with KIM6+ pMcherry from frozen stock.
2. Incubate overnight at 28 °C without aeration.
3. Use 1 ml from this overnight culture to inoculate 100 ml of fresh BHI broth containing 100 μg/ml carbenicillin.
4. Incubate overnight (approximately 18 h) at 37 °C without aeration.
5. Harvest bacteria by centrifugation, resuspend pellet in 1 ml BHI, and store on ice.
6. Prepare the artificial flea feeding device as described by Wade and Georgi [11], Hinnebusch et al. [13], and Bland et al. [12]. Briefly, affix a mouse hide to the glass feeding device with rubber bands and remove hair from the skin using an electric shaver. Attach hosing from 40 °C water bath, through a peristaltic pump to the feeding device. Attach the return hosing to the water bath. Turn on pump to bring the feeding device up to approximately 37 °C. Add 5 ml sterile heparinized mouse blood to the feeding device. Inoculate blood to approximately $1 \times 10^9$ KIM6+ pMcherry bacteria/ml by adding 1 ml bacterial suspension from **step 5** (*see* **Note 1**). Collect approximately 300 *X. cheopis* fleas that have been starved for 4 days, chill fleas on ice, transfer fleas to plastic capsule, and affix to mouse skin side of the glass feeding device using tape. Allow fleas to feed for a minimum of 1 h. Collect all fleas in a 50 ml conical tube, chill on ice and use a dissecting microscope to select fleas that have fed, based on the presence of fresh blood in the digestive tract.

7. House the fed fleas in plastic flea capsules at 21 °C, 75% relative humidity. Provide the fleas with maintenance feedings on mice or artificial feeder every 3 to 4 days.
8. Assess flea blockage immediately after maintenance feeds. Collect fleas into plastic tube and place tube on ice until fleas are immobilized. Use a dissecting microscope with high quality optics placed over a chill table to identify fleas that have fresh, bright red blood in their esophagus, but not in their midgut, indicating that the proventricular valve has been "blocked" by a *Y. pestis* biofilm [12].
9. Store these blocked fleas in a plastic flea capsule overnight at 21 °C, 75% humidity. Use these blocked fleas for transmission experiments within 24 h of blockage diagnosis.

### 3.2 Feeding of Infected Fleas on Mouse Ears

1. Immobilize the blocked fleas by chilling on ice.
2. Place 1 to 10 fleas in custom plastic clip-on feeding chamber (Fig. 1a) (*see* **Note 2**). Store the chamber on ice until ready to feed fleas on a mouse ear.
3. Anesthetize the mouse to be used for intravital microscopy by subcutaneous injection (s.c.) of the ketamine/xylazine mixture (*see* **Note 3**).
4. To facilitate the later identification of fleabites on the mouse ear, inject the mouse intraperitoneally with 100 μl of Sytox Blue in PBS within ten to 20 min of flea feeding (*see* **Note 4**). Sytox Blue is a noncell permeant DNA stain that will stain the nuclei of cells at the bite site due to the dermal cell membrane damage caused by the flea mouthparts.
5. When the mouse is sufficiently anesthetized, place it on an electric heating pad on lowest setting to maintain body temperature (approximately 30 °C). Apply the clip-on feeding chamber to one of the mouse's ears (Fig. 1b). As the tube warms to room temperature, the fleas will become active and attempt to feed. Allow fleas access to the mouse ear for at least 20 min, ideally closer to 40 min (*see* **Note 5**).
6. To terminate feeding, place entire mouse in a chamber containing ~2% isoflurane. Fleas will be anesthetized quickly upon isoflurane exposure. Remove clip-on chamber and collect fleas from ear surface.
7. The mouse can be returned to a temporary cage while the fleas are checked for feeding.
8. Examine the anesthetized fleas using a dissecting microscope to determine the number of fleas that fed. The presence of fresh, bright red blood in the esophagus indicates a feeding attempt.

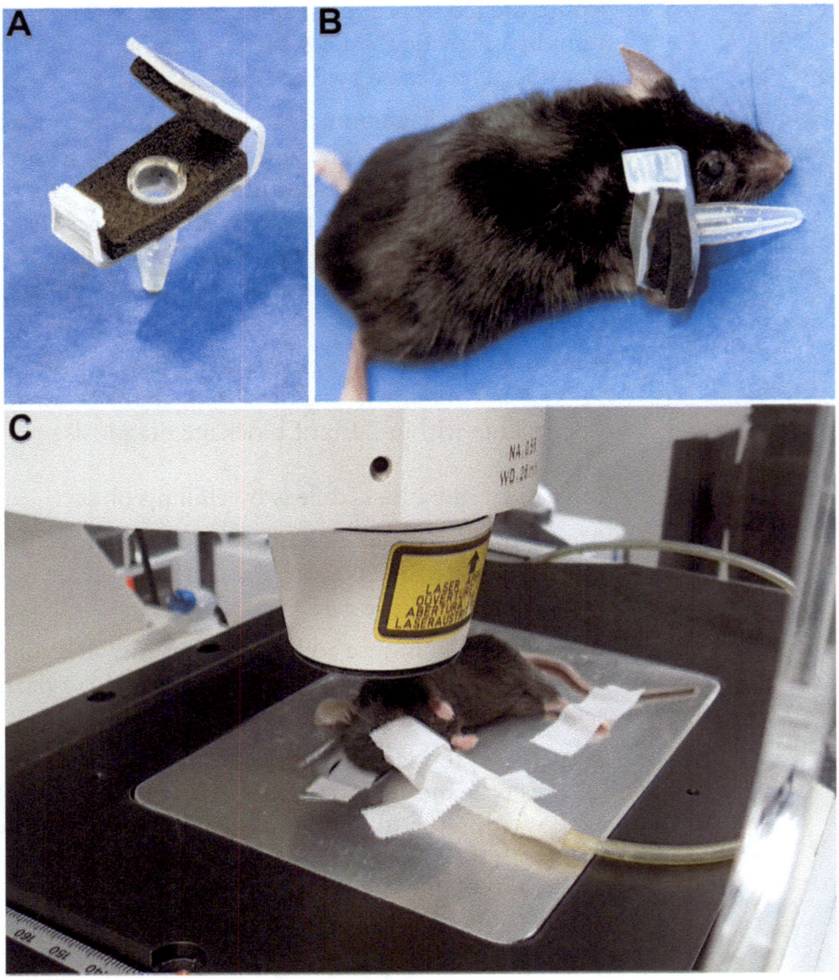


**Fig. 1** Clip-on device for feeding fleas on mouse ears and microscope stage insert set up. (**a**) The clip-on device is constructed using a 200 μl clear PCR tube, adhesive backed foam padding, and plastic cut from the lid of a pipette tip box. Several small holes are pierced in the tube using a 30-gauge needle to allow for air circulation. (**b**) Fleas are placed in the device before it is clipped onto the ventral surface of the ear of a ketamine/xylazine-anesthetized mouse. (**c**) After flea feeding, the mouse is switched to isoflurane anesthesia delivered via nosecone. The ear is gently pressed and spread out, ventral side down, onto a coverslip in the metal stage insert and secured in place with adhesive tape. The hind legs and nosecone tubing are secured to the stage insert with adhesive tape. The stage insert is placed within the incubated chamber and secured into the stage. Figure is reprinted from Shannon et al., *PLOS Pathogens* [9]

### *3.3 Preparation of a Mouse for Imaging by Intravital Confocal Microscopy*

1. Once it has been confirmed that at least one blocked flea has fed on the mouse ear, anesthetize the mouse by placing it in the induction chamber of an Isoflurane vaporizer anesthesia apparatus set to administer 3–4% isoflurane in oxygen (*see* **Note 6**). When the mouse has reached a deep plane of anesthesia, it can be transferred to a breathing circuit with nosecone and the isoflurane can be reduced to 1.5–2% (*see* **Note 7**).

2. Attach a 60 mm × 24 mm coverslip to the microscope stage insert. Move the anesthetized mouse to the stage insert with the ear to be imaged facing down toward the coverslip. Secure the isoflurane tubing and nosecone to the stage insert with adhesive tape so that the nosecone remains over the mouse's nose.
3. Using a cotton swab, thoroughly moisten the ventral surface of the ear to be imaged with sterile dd$H_2O$ and gently place the ear against the coverslip. Use a dry cotton swab to smooth the ear out and ensure as much of the ear as possible is in contact with the coverslip. Use a piece of medical adhesive tape over the ear to secure it to the coverslip. Immobilize the mouse's legs to the stage insert using additional strips of tape (Fig. 1c).
4. Place the stage insert into the microscope stage. Set the incubated chamber enclosing the microscope to 30 °C (*see* **Note 8**).

### 3.4 Identifying and Imaging Flea Bite Sites in the Mouse Ear

1. Using a 10× or 20× objective and an epifluorescence filter set compatible with the 444/480 nm excitation/emission wavelengths of Sytox Blue, the ear can be visually scanned for areas of tissue damage indicative of a flea feeding (*see* **Note 9**). Nuclei of cells damaged by a flea's mouthparts probing the skin will stain bright blue. The staining pattern will vary from a single stab of the mouthparts into the skin (as shown in Fig. 2), to what appears to be the result of multiple probes of the same area (not shown). If the mouse was also injected intravascularly (i.v.) with a vascular dye such as Qtracker655 prior to flea feeding, vascular leakage in close proximity to the Sytox Blue staining confirms the location is a flea bite site (Fig. 2).
2. Once a fleabite has been identified, use the software associated with the particular confocal microscope system to collect a single *Z*-stack containing the Sytox Blue, GFP, mCherry (*see* **Note 10**), and Far Red (if used) emission channels. Parameters such as laser voltage, gain, confocal pinhole size, and *Z*-step size will require optimization for the specific microscope and fluorophores used.
3. Once the various parameters have been optimized, use the software to collect a time series of *Z*-stacks at the desired interval and duration (*see* **Note 11**).
4. An example of the type of images of flea-transmitted *Y. pestis* obtained by this method is shown in Fig. 3. Blocked fleas were allowed to feed on the mouse ear and a flea bite site was identified by Sytox Blue staining. This particular flea bite deposited approximately hundreds of *Y. pestis*. Neutrophil recruitment over the first 1.5 h of imaging can be seen when compared to the 0 h time point (Fig. 3, lower panels).

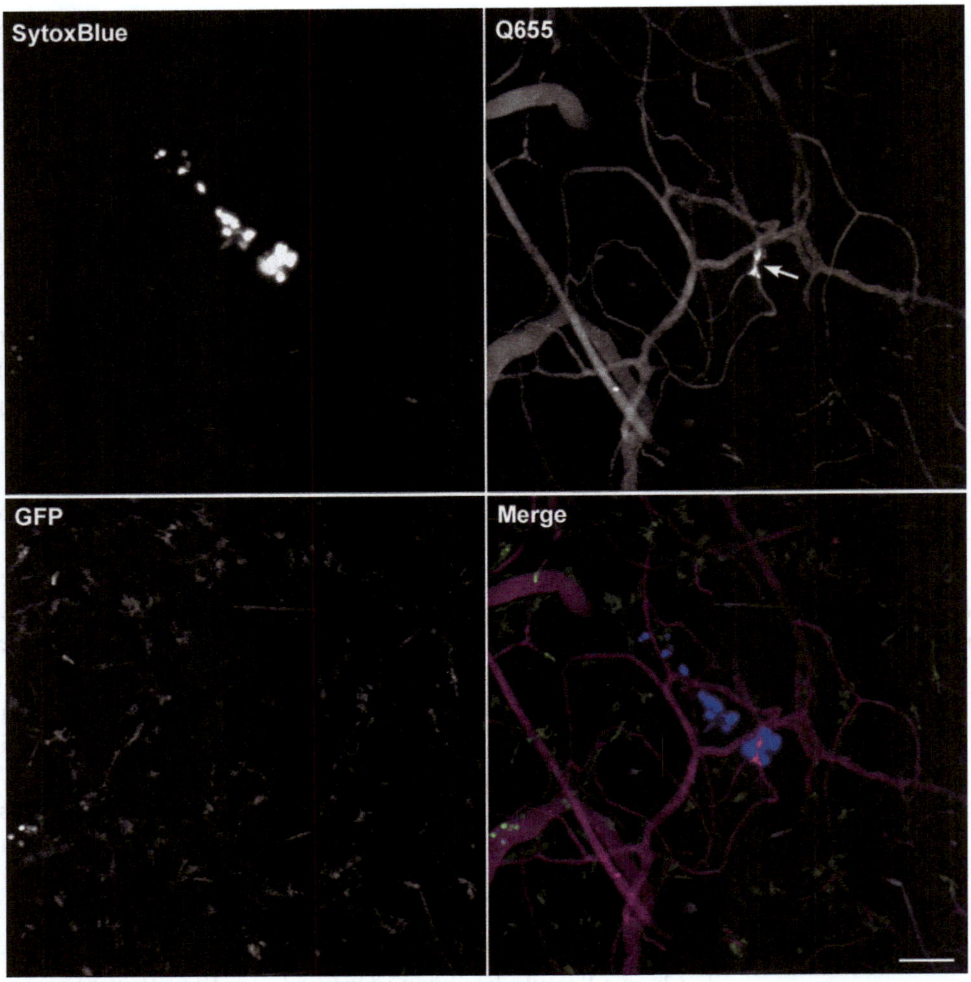


**Fig. 2** Use of the fluorescent DNA stain Sytox Blue to identify uninfected flea bite sites on mouse ears. Confocal images of the ear of a LysozymeM-eGFP mouse injected intraperitoneally (i.p.) with Sytox Blue and i.v. with Qtracker 655 vascular dye prior to being fed upon by uninfected flea. Imaris software was used to make maximum intensity projections of a *Z*-stack for each individual channel (blue, far red [shown as magenta in merge], and green) and a merged image. Nuclei of skin cells damaged by a flea's mouthparts during feeding stain brightly with Sytox Blue. The arrow in the Q655 image indicates leakage of the dye from a blood vessel damaged during flea feeding. A few bright green neutrophils can be seen within a blood vessel at the left side of the image in the GFP channel. The GFP positive, but dim cells are likely tissue-resident macrophages. Scale bar equals 70 μm

### *3.5 Image Processing and Analysis*

1. The image files can be opened within software platforms like Imaris, where the results can be converted to maximum intensity projections or three dimensional still images, montages or movie files suitable for presentation and publication (Fig. 3).
2. The Imaris software can also quantify parameters like bacteria-neutrophil colocalization/interaction, or neutrophil recruitment, migration speed, and directionality as described in the developer's instructions [9] (*see* **Note 12**).

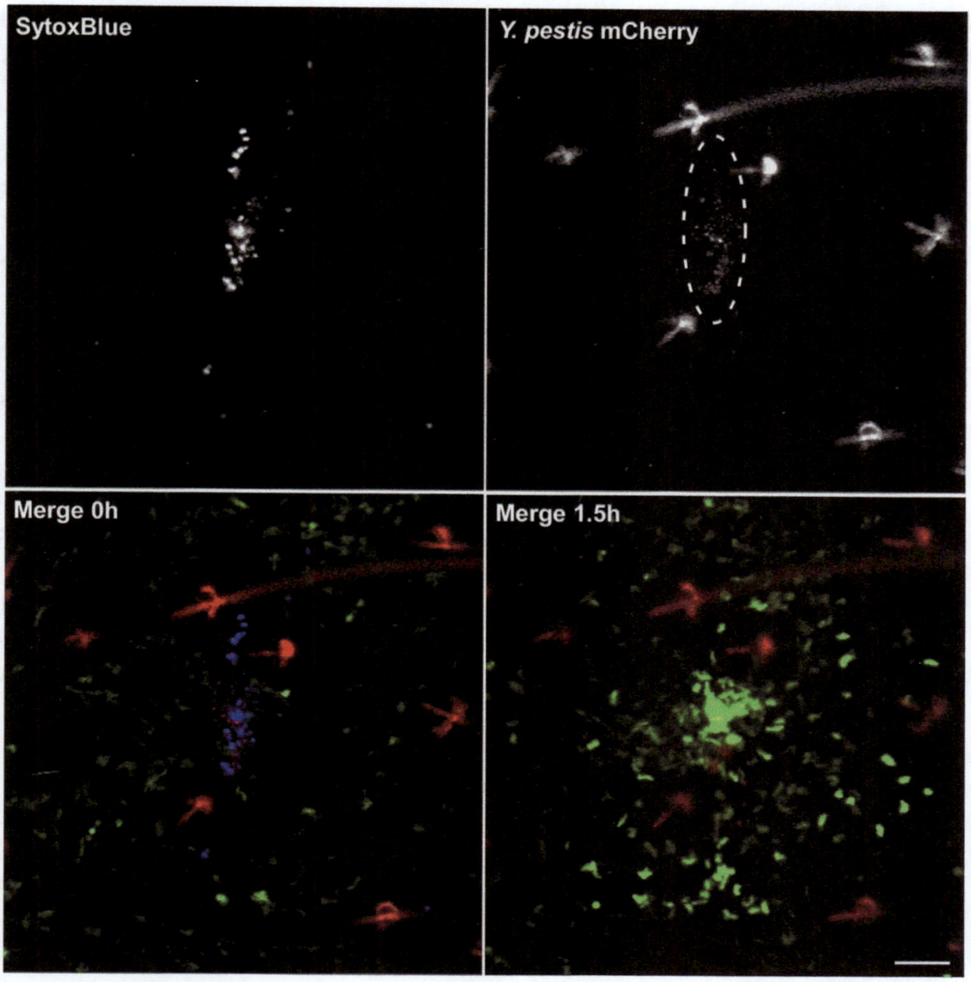


**Fig. 3** Imaging neutrophils at a flea bite site containing flea-transmitted *Y. pestis*. Confocal images of the ear of a LysozymeM-eGFP mouse injected i.p. with Sytox Blue prior to being fed upon by fleas infected (blocked) with *Y. pestis* expressing mCherry. Imaris software was used to make maximum intensity projections of a *Z*-stack for the Sytox Blue and mCherry channels (upper panels). Merged images of the blue, red and green (GFP) channels show the 0 h and 1.5 h time points (lower panels). This flea bite site was identified by the Sytox Blue staining. The dashed ellipse in the red channel indicates an area where *Y. pestis* were deposited in the skin (note that hair and hair follicles autofluoresce in the red channel). Only a few GFP-bright neutrophils can be seen in the green channel at 0 h; large numbers of neutrophils were recruited to this bite site by the 1.5 h time point. Scale bar equals 70 μm

## 4 Notes

1. Following the procedure as described in **steps 5** and **6** typically results in bacterial concentrations in the infectious blood meal ~$1 \times 10^9$/ml. A small aliquot of blood should be set aside, diluted, and plated on blood agar plates to determine actual concentration of colony-forming units in the blood meal.

2. We construct this clip-on feeding device by gluing together a 200 μl clear PCR tube, adhesive-backed foam padding, and some plastic cut from the lid of a pipette tip box (Fig. 1a). Alternatively, we have had success cutting the tip off of a 10 ml disposable syringe and gluing a fine mesh fabric, with openings large enough for fleas to feed through, across the opening [14]. The mesh end of this syringe feeder can be pressed against the ear of the anesthetized mouse allowing the fleas to feed.
3. Ketamine/xylazine mixture is used for anesthesia because fleas are highly sensitive to isoflurane, and use of isoflurane delivered to the mouse by nosecone can result in inadvertent anesthesia of the fleas placed on the ear. Inject 100 μl/20 g mouse weight of ketamine/xylazine mixture s.c.
4. When first establishing this method, i.v. injection of a vascular dye like Qtracker 655 prior to flea feeding can help confirm that the Sytox Blue is staining flea bite sites. Qtracker 655 will allow visualization of the vasculature in the far red fluorescence channel and small areas of blood vessel damage can be identified by leakage of the dye. Flea feeding will result in a small amount of vascular leakage at the bite site as can be seen in Fig. 2.
5. This method typically results in a plane of anesthesia deep enough to permit flea feeding for a minimum of 40 min. Blocked fleas are unable to ingest a blood meal and will become weak and dehydrated, and thus can be more reluctant to feed than uninfected fleas. Allowing 4 to 5 blocked fleas access to the mouse ear for 40 min may be required to ensure at least 1 flea will attempt to feed.
6. For initial experiments, it may be better to use uninfected fleas to verify that flea bites can be identified using the Sytox Blue method.
7. Periodically monitor respiration rate/effort and pedal reflex to evaluate depth of anesthesia according to IACUC guidelines.
8. If the microscope does not have an incubated chamber, a heated stage insert may be adapted for maintaining mouse body temperature. If neither is available, the heating pad used during flea feeding can be placed over the mouse on the stage.
9. Sytox Blue fluorescence can be visualized with filter sets used for DNA stains like DAPI or Hoescht; however, these filters typically are suboptimal for the peak emission (480 nm) of Sytox Blue. Use of a filter set specifically suited for this dye will improve results.
10. Small hairs on the surface of the mouse ear as well as hair follicles will autofluoresce in multiple channels, but predominantly in the red channel, as is evident in Fig. 3.

11. The interval between scans will be determined by the cell type being studied and the types of post-acquisition analyses to be applied to the images. Scan intervals between one and two minutes are typically sufficient to track neutrophil accumulation and interaction with bacteria. Shorter intervals may be required to trace the movement of individual cells over time.
12. Tutorials available as part of these software packages are highly informative and cover quantitative image analyses beyond the scope of what can be discussed here.

## Acknowledgements

We thank Clayton Jarrett, Chris Bosio, Dave Bland, Dustin Van Hofwegen, Adelaide Miarinjara, and Ashley Schwarzer for their careful review of the manuscript, David Dorward and Vinod Nair for assistance with confocal microscopy, and Anita Mora for assistance with graphic arts. This research was supported by the Intramural Research Program of the National Institutes of Health, National Institute of Allergy and Infectious Diseases.

### References

1. Jain R, Weninger W (2013) Shedding light on cutaneous innate immune responses: the intravital microscopy approach. Immunol Cell Biol 91(4):263–270. https://doi.org/10.1038/icb.2012.76
2. Faust N, Varas F, Kelly LM, Heck S, Graf T (2000) Insertion of enhanced green fluorescent protein into the lysozyme gene creates mice with green fluorescent granulocytes and macrophages. Blood 96(2):719–726
3. Burnett SH, Kershen EJ, Zhang J, Zeng L, Straley SC, Kaplan AM, Cohen DA (2004) Conditional macrophage ablation in transgenic mice expressing a Fas-based suicide gene. J Leukoc Biol 75(4):612–623. https://doi.org/10.1189/jlb.0903442
4. Jung S, Aliberti J, Graemmel P, Sunshine MJ, Kreutzberg GW, Sher A, Littman DR (2000) Analysis of fractalkine receptor CX(3)CR1 function by targeted deletion and green fluorescent protein reporter gene insertion. Mol Cell Biol 20(11):4106–4114
5. Lindquist RL, Shakhar G, Dudziak D, Wardemann H, Eisenreich T, Dustin ML, Nussenzweig MC (2004) Visualizing dendritic cell networks in vivo. Nat Immunol 5 (12):1243–1250. https://doi.org/10.1038/ni1139
6. Unutmaz D, Xiang W, Sunshine MJ, Campbell J, Butcher E, Littman DR (2000) The primate lentiviral receptor Bonzo/STRL33 is coordinately regulated with CCR5 and its expression pattern is conserved between human and mouse. J Immunol 165 (6):3284–3292
7. Benson RA, Brewer JM, Garside P (2017) Visualizing and tracking T cell motility in vivo. Methods Mol Biol 1591:27–41. https://doi.org/10.1007/978-1-4939-6931-9_3
8. McArdle S, Mikulski Z, Ley K (2016) Live cell imaging to understand monocyte, macrophage, and dendritic cell function in atherosclerosis. J Exp Med 213(7):1117–1131. https://doi.org/10.1084/jem.20151885
9. Shannon JG, Bosio CF, Hinnebusch BJ (2015) Dermal neutrophil, macrophage and dendritic cell responses to *Yersinia pestis* transmitted by fleas. PLoS Pathog 11(3):e1004734. https://doi.org/10.1371/journal.ppat.1004734
10. Shannon JG, Hasenkrug AM, Dorward DW, Nair V, Carmody AB, Hinnebusch BJ (2013) *Yersinia pestis* subverts the dermal neutrophil response in a mouse model of bubonic plague. MBio 4(5):e00170–e00113. https://doi.org/10.1128/mBio.00170-13

11. Wade SE, Georgi JR (1988) Survival and reproduction of artificially fed cat fleas, *Ctenocephalides felis* Bouché (Siphonaptera: Pulicidae). J Med Entomol 25:186–190
12. Bland DM, Brown LD, Jarrett CO, Hinnebusch BJ, Macaluso KR (2017) Methods in flea research. Biodefense and emerging infections research resources repository. https://www.beiresources.org/Catalog/VectorResources.aspx
13. Hinnebusch BJ, Perry RD, Schwan TG (1996) Role of the *Yersinia pestis* hemin storage (*hms*) locus in the transmission of plague by fleas. Science 273(5273):367–370
14. Bosio CF, Viall AK, Jarrett CO, Gardner D, Rood MP, Hinnebusch BJ (2014) Evaluation of the murine immune response to *Xenopsylla cheopis* flea saliva and its effect on transmission of *Yersinia pestis*. PLoS Negl Trop Dis 8(9): e3196. https://doi.org/10.1371/journal.pntd.0003196

# Chapter 6

# Laser Scanning Microscopy of *Yersinia pestis* Infected Tissues

**Rodrigo J. Gonzalez**

## Abstract

Laser scanning microscopy (LSM) is a technology that allows for direct observations of host-pathogen interactions during infection. Two of the most available forms of LSM are confocal and two-photon LSM. In addition to high resolution and contrast, these two technologies also provide high excitation penetrance in unsectioned samples. High penetrance allows for imaging of layers of tissue that are difficult to image with other more conventional microscopy approaches. Thus, confocal and two-photon LSM open the possibility of observing infection in a three-dimensional context, where the natural architecture of a tissue is preserved. Few studies have used LSM technology to gain insights into *Yersinia pestis* pathogenesis in the mammalian host. The use of LSM in the plague field has an enormous potential for the discovery of the mechanisms that lie behind key aspects of pathogenesis such as colonization, dissemination, and tissue damage. This chapter provides guidance for the implementation of confocal or two-photon LSM to study *Y. pestis* interactions with the host in unsectioned tissues. This document provides specific instructions applied to imaging of *Y. pestis*, and also discusses relevant aspects of imaging, such as the operation of laser scanning microscopes and the use of fluorescent probes.

**Key words** Laser scanning fluorescence microscopy, Microscopy imaging, Two-photon microscopy, Confocal microscopy, Ear pinna, Skin, Plague, *Yersinia pestis*

## 1 Introduction

Microscopy imaging is a technique that has influenced the study of infection tremendously, as it can truly provide direct observations of host-pathogen interactions in the tissues they occur. Laser scanning microscopy (LSM) in particular is an approach that has proven extremely valuable for the study of microbial pathogenesis, as it allows for high optical resolution, augmented imaging contrast, and more importantly, the acquisition of images from unsectioned tissues [1]. The analysis of unsectioned tissues is a very desirable feature as it allows for studies of infection in a three-dimensional context. Consequently, LSM opens new possibilities for the understanding of how pathogens interact with the tissues they infect. Only a few studies have incorporated LSM imaging to investigate

Viveka Vadyvaloo and Matthew B. Lawrenz (eds.), *Pathogenic Yersinia: Methods and Protocols*, Methods in Molecular Biology, vol. 2010, https://doi.org/10.1007/978-1-4939-9541-7_6, © Springer Science+Business Media, LLC, part of Springer Nature 2019

plague pathogenesis in mice [2–4]. Possible reasons are the lack of equipment and, perhaps, more commonly, expertise to operate it. However, at present, most institutions offer facilities that provide access to LSM equipment, in addition to highly trained personnel to provide instruction on how to operate it. Moreover, in the case of fully virulent *Yersinia pestis*, fixed samples can be processed and imaged outside biosafety level three (BSL3) facilities. The purpose of this chapter is to serve as a guide for investigators with little or no experience in LSM and who want to incorporate this technique into their research.

Confocal and two-photon LSM allow for the imaging of unsectioned tissues and are the two most accessible alternatives for the imaging of *Y. pestis* infected samples. In confocal LSM, a thick sample is excited by a cone of light at multiple planes [1]. Excitation by this cone at planes other than the plane of acquisition generates undesired background signal, a problem that is corrected with the use of a pinhole. The use of this pinhole excludes background signal, making it possible to obtain high contrast images. In two-photon LSM, the sample is not excited by a cone but by a single excitation point that excites only at the level of the plane of acquisition [1, 5]. Thus, no background signal is generated. Moreover, two-photon LSM allows for red-shifted excitation, which results in higher penetrance to image much deeper into a tissue in comparison to confocal LSM [5, 6].

Confocal and two-photon LSM technologies acquire signal from thick samples through single two-dimensional optical slices (*see* Fig. 1). A single optical slice generates a cross-section image of the sample. As multiple optical slices are collected through the *Z* axis, an optical stack is obtained. Optical stacks can be processed by visualization software to render a tridimensional image for further analysis such as identification or quantification of cellular components of the sample, physical interactions between these components, etc. [1]. While the approach to be used (i.e., confocal LSM vs. two-photon LSM) will determine how deep into the tissue acquisition can take place, the sample itself is also of importance, as different tissues have different light scattering properties. Recently developed clearing techniques [7–9], which are not discussed in this chapter, can improve dramatically the ability of LSM to obtain images from very deep layers of a tissue.

Visualization through LSM can only occur if the cell or tissue under study is labeled with a fluorophore. In the case of *Y. pestis*, the best alternative is the generation of strains that can synthesize a fluorescent protein [10]. Visualization through the use of fluorescently labeled antibodies is not recommended because it can generate unspecific signal. Green fluorescent protein (GFP) [2] and red fluorescent protein (RFP), as DsRed[2] or mCherry[3], have been used successfully to image *Y. pestis* through LSM in mouse tissues. In the case of the host, while mice that express a fluorescent protein

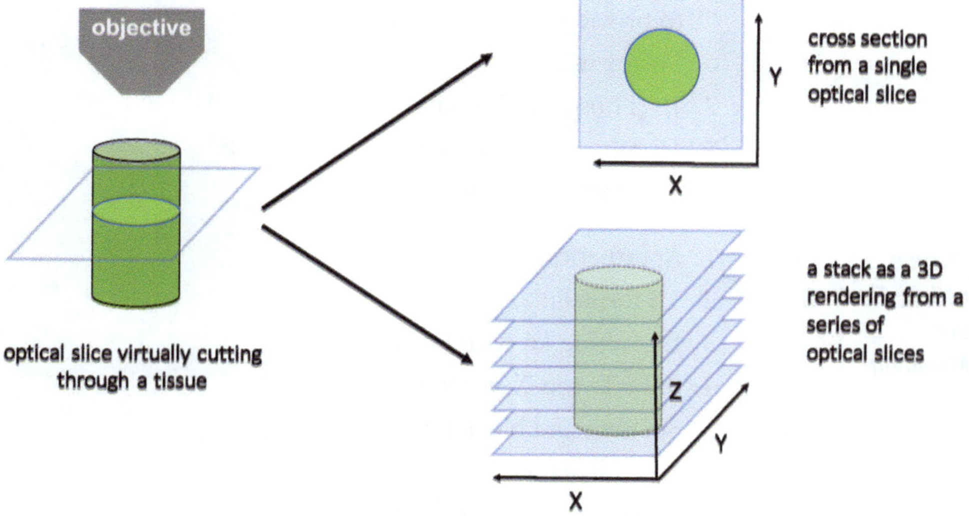


**Fig. 1** Diagram showing an optical slice virtually cutting into a sample (represented as a green cylinder). A single optical slice can generate a cross section of the sample (top right). A series of optical slices, an optical stack, can recreate a tridimensional image of the sample (bottom right)

in the tissues or cells of interest are preferred [11], they are not always available and the use of staining antibodies has to be considered.

The opportunity to make direct observations of infection as it develops in tissues is a fascinating possibility in plague research. LSM is a technology through which this possibility can be accomplished. The present chapter describes procedures for the imaging of fixed unsectioned tissues of mice infected with *Y. pestis*. The chapter begins by providing guidance on how to determine attenuation of *Y. pestis* that express a fluorescent protein. Assessing attenuation of fluorescently labeled pathogens should be the foundation of any imaging experiment whose goal is to describe phenomena that occur during natural infections. In addition, procedures to harvest, fix, stain, and mount tissues to render them suitable for imaging are also discussed. Lastly, the chapter describes basic procedures to acquire good quality images and to conduct post-acquisition adjustments to these images. While all procedures focus on *Y. pestis* research, they are written to be easily adapted to the use of other pathogens.

## 2 Materials

### 2.1 Testing for Growth Defects and Attenuation of Fluorescently Labeled *Y. pestis*

1. Brain heart infusion (BHI) agar plates.
2. Inoculation loops.
3. BHI broth.
4. Dissecting microscope with fluorescence light source.

5. Microbial culture rotator (i.e., microbial culture roller drum) to provide aeration to cultures.
6. Vortex mixer.
7. 1× Phosphate-buffered saline (1× PBS)
8. 15 mL conical tubes for serial dilutions
9. Incubator at 26 °C.

#### 2.2 Tissue Harvesting, Fixation, Staining, and Mounting

1. Anesthetics.
2. 4% formalin.
3. Dissecting kit with fine tweezers.
4. 0.3% Triton-X in 1× PBS.
5. Staining and blocking antibodies.
6. Nuclear stain.
7. Glass slides and coverslips.
8. Hardening mounting medium.

#### 2.3 Imaging and Analysis

1. Laser scanning microscope (i.e., confocal or two-photon excitation).
2. Software for imaging analysis (e.g., Fiji, Imaris, etc.).

## 3 Methods

For the following section, it is assumed any procedures involving fully virulent *Y. pestis* will be conducted in a biosafety cabinet in a BSL3 facility and will follow established guidelines for select agents.

#### 3.1 Testing for Growth Defects of Fluorescently Labeled Y. pestis In Vitro

1. Streak a plate of BHI agar with wild type (WT) nonfluorescent (hereinafter referred to as WT) *Y. pestis* from a frozen stock culture kept at −80 °C.
2. Repeat the same procedure with a frozen stock culture of fluorescently labeled *Y. pestis*, using one BHI agar plate with no antibiotics and one BHI agar plate with the antibiotics the fluorescently labeled strain is resistant to.
3. Incubate the three plates at 26 °C for 48 h.
4. Observe individual colonies and determine if bacterial growth (i.e., colony size) and overall colony characteristics (i.e., shape, etc.) in the three plates are comparable (*see* **Note 1**).
5. Compare isolated colonies of the fluorescently labeled strain growing on plates with and without antibiotics under a dissecting microscope equipped with a fluorescence light source (*see* **Note 2**).

6. With an inoculation loop, scrape off WT *Y. pestis* from the area of the plate with most confluent growth (typically the very first lines streaked with the inoculation loop when streaking from a frozen stock culture) and add the bacteria to a glass test tube with 2 mL BHI broth.
7. Repeat the same procedure with the fluorescently labeled strain (from the plate of BHI agar with antibiotics), using BHI broth with the appropriate antibiotics.
8. Incubate both tubes at 26 °C with aeration (e.g., use a microbial culture roller drum) for approximately 15 h.
9. Compare the cloudiness of the two tubes, as well as the formation of the characteristic biofilm that forms a ring on the wall of the tube (*see* **Note 3**).
10. Take the tube with the culture of fluorescently labeled bacteria and mix vigorously with a vortex mixer to integrate the biofilm into the rest of the culture.
11. Make serial dilutions to obtain individual colonies (as opposed to a lawn) when plating. Plate on BHI agar plates with and without antibiotics.
12. Incubate the plates at 26 °C until well-defined colonies are visible (approximately 48 h).
13. Count the number of colonies on the plate with and without antibiotics for the same dilution (*see* **Notes 4** and **5**).

### 3.2 Testing Attenuation of Fluorescently Labeled *Y. pestis* In Vivo

1. Inoculate two groups of mice: one with WT and the other with fluorescently labeled *Y. pestis*, using a route of inoculation and inoculum relevant for the overall goals of the experiment that will be conducted ultimately.
2. Assess for signs of disease such as hunching, ruffled fur, low responsiveness, etc., at appropriate time points (i.e., usually time points when systemic dissemination already took place) in each group.
3. Humanely sacrifice the mice at relevant time points for the selected infection model (e.g., intradermal, etc.) and harvest the tissues of interest (i.e., lymph nodes, spleens, skin, etc., *see* **Note 6**).
4. Homogenize the tissues.
5. Plate the homogenized tissues (using the appropriate dilutions to obtain single colonies) from both strains on BHI agar with no antibiotics, and in the case of the fluorescent strain, plate on BHI agar with the appropriate antibiotics as well. Plating the fluorescently labeled strain with no antibiotics will give insights into the stability of the insert in vivo.

6. Incubate the plates at 26 °C until single colonies are visualized (approximately 48 h).
7. Count the number of CFU per tissue (per group) and establish differences between groups (*see* **Notes** 7 and **8**).

### 3.3 Tissue Preparation and Imaging

#### 3.3.1 Tissue Harvesting, Fixation, and Staining

1. Inoculate mice with fluorescently labeled *Y. pestis* using a route of inoculation and inoculum that are relevant for the overall goals of the experiment (*see* **Note 9**).
2. Inject any dye or fluorescently labeled antibody to the infected mice using a relevant route (*see* **Note 10**).
3. Humanely sacrifice the animal at a relevant time point (*see* **Note 11**).
4. Collect the desired tissues and fix overnight at 4 °C in 4% formalin (*see* **Notes 12** and **13**).
5. At this point, the tissues can be removed from the BSL3 facility (*see* **Note 14**).
6. Wash the tissues in 1× PBS at 4 °C in a rotator for an hour, twice, to remove any excess formalin.
7. If an intradermal model of inoculation in the ear pinna is used, consider separating the dorsal and ventral leaflets of the ear (*see* **Note 15** and Fig. 2).
8. Transfer to a microcentrifuge tube with 0.3% Triton X-100 in 1× PBS.
9. If needed, add blocking antibodies to avoid unspecific signal.
10. Incubate overnight with the blocking antibodies at 4 °C in a rotator, making sure the liquid bathes the entire tissue.
11. Replace the liquid with 0.3% Triton X-100 PBS and staining antibodies (*see* **Note 16**) with the desired fluorophore (*see* **Note 17**) and at the appropriate concentration (*see* **Note 18**).
12. If necessary, add nuclear stain to the staining cocktail (*see* **Note 19**).
13. Incubate overnight (*see* **Note 20**) at 4 °C in a rotator making sure the staining cocktail bathes the entire tissue.
14. Wash with 0.3% Triton-X 100 in 1× PBS for approximately 2 h.
15. Place the tissue on a glass slide and remove excess liquid with an absorbent pad.
16. Add hardening mounting medium (*see* **Note 21**), remove any bubbles that might have formed, and place a coverslip on top.
17. Let the mounting medium dry as specified by the provider (*see* **Note 22**).

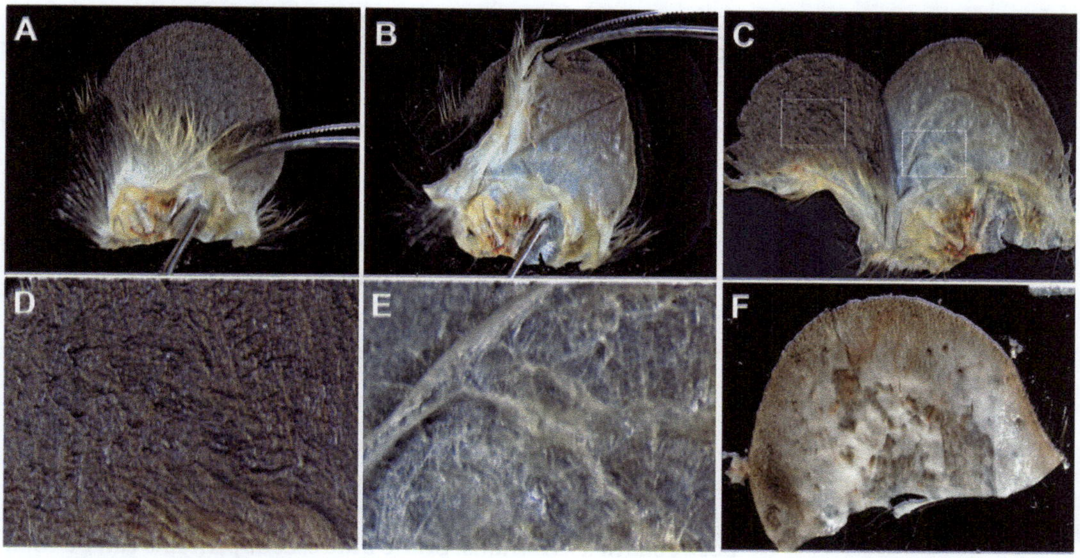


**Fig. 2** Separation of the two leaflets of a fixed ear pinna before staining. (**a**) Two layers of skin at the base of the ear being held with forceps. (**b**) Separation of the two leaflets of the ear. (**c**) The two leaflets already separated. The dorsal leaflet is on the left with no cartilage on the surface, and the ventral leaflet is on the right, with a layer of cartilage attached to it. The dotted boxes depict magnified areas shown in **d** (left side) and **e** (right side) to show absence and presence of cartilage, respectively. (**f**) A dorsal leaflet separated from the ventral leaflet in an incorrect manner, as pieces of cartilage remained attached. Undesired cartilage can be removed with fine tweezers, but the procedure is very time-consuming and requires extreme care

#### *3.3.2 Imaging and Analysis*

1. Place the sample under the microscope and select the desired magnification (*see* **Note 23**) to survey the tissues (*see* **Note 24**).
2. Once the tissue has been surveyed, define the area that will be imaged.
3. Determine if a single slice or a stack will be acquired (*see* **Note 25**).
4. If a stack will be acquired, define its thickness (*see* **Note 26**), making sure signal is not compromised at the deepest point of acquisition (*see* **Note 27**).
5. Define step size. This is, the distance between each optical slice in the stack (*see* **Note 28**).
6. Define pixel dwell time, the time the laser illuminates a single point or pixel (*see* **Note 29**).
7. Adjust laser power and photomultiplier (PMT) gain settings (*see* **Note 30**).
8. Start acquisition and wait until the entire stack has been acquired.
9. Transfer the acquired dataset to the visualization and analysis software of preference (*see* **Note 31**).

10. Adjust signal from every channel by selecting only one of the acquisition channels at a time and adjusting the settings for brightness and contrast (*see* **Note 32**).
11. Verify that the values shown in the scale bar are correct (*see* **Note 33**).
12. Conduct quantification analysis if needed.
13. Determine the best presentation format for the question being addressed by the image (*see* **Note 34**).

## 4 Notes

1. While ideally a fluorescently labeled strain should not behave differently from a WT strain, this is not always the case. Thus, it is important to determine if there are any differences in growth between the two strains. Some fluorescent proteins can be toxic as they accumulate in the bacterial cell. In other cases, the production of a fluorescent protein under a strong promoter, for instance, results in high energy expenditure by the bacteria. This might divert resources normally allocated for growth. If severe growth defects are identified, the use of a different fluorescent protein or promoter might be considered.
2. This step will determine how stable the insert containing the fluorescent protein is in *Y. pestis*. If expression of the fluorescent protein has a high metabolic cost, then the insert could be lost. This is most likely to occur if the sequence encoding the fluorescent protein exists in a plasmid rather than in the bacterial chromosome. In the absence of antibiotics, in the context of an energetically demanding (or toxic) fluorescent protein, bacteria that do not inherit the plasmid encoding the protein will be selected, as they will grow more efficiently than those that inherited the plasmid.
3. The goal of this step is to determine whether or not two aspects of normal growth (i.e., cloudiness in the tube and formation of the biofilm ring) are not affected in the fluorescently labeled strain. Cloudiness could be affected, for instance, by bacteria not growing at a normal rate, or clumping together in the bottom of the tube as viability is lost. The formation of the biofilm ring could be affected if the insert compromises the bacteria's ability to clump together as a biofilm, for instance.
4. The number of colonies on each plate (with and without antibiotics) should be comparable if the genetic insert is not being lost. The presence of the fluorescent protein can also be assessed on both plates with a fluorescence dissecting microscope, as described in **step 5** of this section.

5. Determining the stability of the sequence encoding the fluorescent protein in vitro is a simple step that should reveal any major problems with the fluorescent strain. Because no major pressure of selection exists in vitro, the insert should be stable in most fluorescent strains. However, insert stability will not necessarily replicate in vivo, and for this reason, more complex experiments using mice must be conducted (*see* next section). If a fluorescent strain already shows problems in vitro, it is very likely it will not work well in vivo. When designing a strategy to visualize bacteria by the expression of a fluorescent protein from a plasmid, consider incorporating a toxin-antitoxin system (e.g., hok/sok system) for plasmid maintenance.
6. Always include a later time point (e.g., 48–72 h post inoculation for an ear intradermal model), as this time point will reveal the stage of bacteria after extensive dissemination. It is at later time points when most immune responses have already been triggered, possibly generating selective pressure on the bacteria.
7. If mice were equally treated (i.e., same inoculum, route of inoculation, and tissues were harvested at the same time point) and the fluorescently labeled strain is not attenuated, CFU counts between WT and fluorescently labeled *Y. pestis* should be comparable. Many reasons can account for lower burdens in mice infected with the fluorescently-labeled strain, the fluorescent protein of choice is one of them. While GFP tends to be extremely bright and, hence, ideal for imaging, strains expressing this protein can show delayed kinetics of infection. RFP-expressing strains, however, have shown the exact same kinetics of infection as a WT strain [2]. In addition to infection kinetics, consider assessing survival times between mice infected with each strain. Differences in the kinetics of infection or survival times between both strains should be disclosed and taken into account when making inferences on the questions the study is addressing.
8. If the results reveal that the fluorescently labeled strain is highly attenuated (e.g., severely delayed kinetics, no signs of disease, no mortality, etc.), consider using a different fluorescent protein or a different promoter. An attenuated fluorescently labeled strain might reveal confounding host-pathogen interactions (e.g., susceptibility to a phagocyte) that are inherent to the strain per se and not to WT *Y. pestis*.
9. If interactions between bacteria and a specific tissue or cell type are to be assessed, consider using genetically engineered mice that express a fluorescent protein in that tissue or cell type [10, 11]. Be aware that some of these systems will show high specificity, while others are rather leaky, showing fluorescence at nonrelevant tissues. Either for fluorescently labeled bacteria

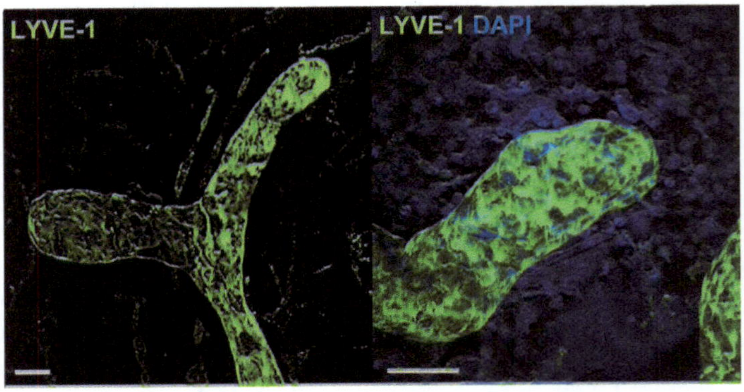


**Fig. 3** Injection of fluorescently labeled anti-LYVE-1 into the ear insterstitium results in suitable staining of terminal lymphatic vessels in the skin. Each image shows termninal lymphatic vessels at different magnifications to appreciate the level of resolution that can be attained. In the absense of nucelar stain (left image), the presence of stained nonlymphatic tissue is evident. Scale bar is 20 μm

or mice, make sure the microscope is equipped with the appropriate emission filters (filters to collect signal) to visualize the fluorescent proteins of choice and that these filters are of the appropriate bandwidth to avoid spectral bleed-through from one channel into another.

10. In some instances, it is necessary to inject a dye or fluorescently labeled antibody directly into the mouse to stain for a specific tissue. For instance, in the case of lymphatic vessels of the ear, the vessels will stain better if fluorescently labeled anti-LIVE-1antibody is injected into the dermis many hours before sacrificing the mouse [2] (*see* Fig. 3). In the case of visualization of neutrophils that extravasate into an infected tissue, fluorescently labeled anti-Ly6G antibody can be injected intravenously[2].

11. Be aware that the kinetics of dissemination and proliferation of *Y. pestis* in the host change significantly, depending on tissue and time point after inoculation. For instance, for WT C57BL/6 mice inoculated intradermally into the ear with approximately 300 CFU, the number of bacteria to be detected in the draining lymph node at 24 h post-inoculation will be so high that imaging might be difficult [2]. In the same model, numbers of *Y. pestis* CFU at the site of inoculation (ear) are expected to be low, as *Y. pestis* does not seem to proliferate nearly as much in this tissue as it does in lymphatic tissues.

12. Formalin is a saturated formaldehyde solution and it is commercially available at different concentrations. Care has to be placed into concentration and time of exposure, since some fluorescent proteins quench when exposed to formalin.

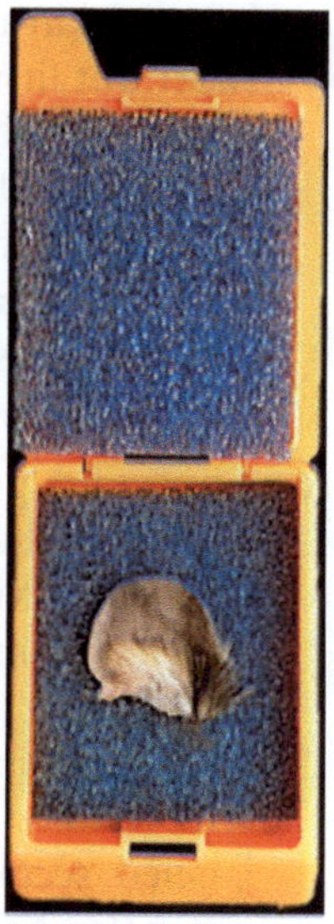

**Fig. 4** Open plastic histology cassette (orange), with sponge pads (blue), depicting an ear about to be fixed. The cassette is closed in such a way that the ear is pressed in between the two sponge pads before submersion into a fixative

13. Certain tissues will need to be fixed in such a way that their shape will be preserved. These tissues can be fixed using a histology cassette with pressing sponge pads (*see* Fig. 4).
14. It is very important to confirm that the conditions that are described in **step 5** are sufficient to kill *Y. pestis* so the tissues can be taken outside the BSL3 facility for staining and imaging. One of the simplest ways to test for bacterial viability after fixation is to rinse the tissue with PBS (to eliminate excess formalin), mince it into particularly small pieces, and incubate it in BHI broth (in a glass test tube, with aeration) at 26 °C for at least 48 h. If the procedure killed the bacteria the broth should look clear after incubation. Remember that thicker organs (e.g. liver and thymus), will take longer for proper fixation than organs that are thin (e.g. ear pinna). Discuss this and other possible procedures to test for viability in advance with the biosafety officer in charge of the BSL3 facility to be used.
15. Because of excitation penetrance, the two leaflets of the ear must be separated before mounting the sample when using confocal microscopy. To successfully separate the leaflets, localize the base of the ear and observe how there are two layers of skin that are separated by connective tissue. These two layers of skin become the ventral and dorsal leaflets of the ear pinna, separated only by cartilage, as one moves away from the base toward the edges of the ear. With forceps, hold each separate

layer of skin at the base of the ear. Pull the two layers apart with short but firm movements. If done correctly, the cartilage will remain attached to the ventral leaflet (assuming injection was done in the dorsal leaflet). The leaflet to be imaged should not have any cartilage attached. Separating the two ear leaflets is a procedure that might not be necessary when using two-photon LSM, as this technology allows for higher penetrance than confocal LSM.

16. Antibodies per se are not visible through fluorescence microscopy, and fluorophores must be bound to them. While many antibodies with fluorophores bound to them are widely available in the market, this is not always the case and other alternatives have to be considered. One of these alternatives is the use of biotinylated antibodies, to which fluorophore-bound streptavidin can bind. Another alternative is fluorescently-labeled secondary antibodies, which will bind to a specific unlabeled primary antibody. Lastly, a labeling kit allows binding a wide range of fluorophores to unlabeled antibodies. For the biotinylated antibody and secondary antibody strategies, two-step staining is required.

17. When selecting a fluorophore, make sure it is suitable for the excitation source (lasers) and emission filters in the microscope. If the appropriate excitation source or emission filter is not available, signal from the fluorophore will not be detected. As previously discussed in **Note 9**, the appropriate emission filters can prevent spectral bleed-through. Moreover, some fluorophores might work under confocal LSM but might not work as well under two-photon LSM or vice versa. The Alexa Fluor dyes are excellent fluorophores for microscopy applications because of their high stability and brightness. In general, these dyes can be used interchangeably for confocal and 2-photon microscopy, but there are exceptions. For instance, Alexa Fluor 647 works well in both technologies, but under two-photon excitation, high levels of photobleaching can be observed. Personnel in charge of microscopy facilities are very familiar with similar problems and it is recommended to assess staining panels with them in advance. Furthermore, use appropriate controls to discern between genuine signal and autofluorescence or to determine that the antibody of preference is specific for the target tissue. Lastly, some molecules are regarded as optimal markers for specific tissues, but these molecules might also exist in other tissues as well. An example is LYVE-1, a good option as a target to visualize lymphatic vessels. Expression of LYVE-1 is mostly seen in superficial lymphatics, but other cells in the skin also express some levels of this protein (*see* Fig. 3).

18. For detection to be successful, the fluorophore has to be applied at an optimal concentration. Too low or too high of a concentration can result in little to no signal or nonspecific signal, respectively. Optimal staining concentrations are typically determined empirically. Many staining antibodies are available at a concentration of approximately 0.5 mg/mL and a rule of thumb is to make a staining cocktail with these antibodies using them in a 1:200 dilution. However, this concentration will not always work and adjustments might need to be made.
19. The use of a nuclear stain brings many advantages to microscopy imaging. When analyzing images, many inferences are made by context, something that a nuclear stain can provide. For instance, the presence of a cell of interest can be confirmed, in addition to shape and size, by the visualization of a nucleus. Furthermore, nuclear stains can also delineate the architecture of a tissue, making different layers of the tissue (e.g., epithelium) visible. DAPI and Hoechst, with emissions around 450 nm (blue), are the most common options for nuclear stains. However, there are many other, perhaps less well known, nuclear stains with different emission wavelengths that can be used if the blue channel is not available.
20. Overnight incubation allows for even staining of a thick piece of tissue or an entire organ (e.g. a spleen). Incubation times can be reduced if the tissues are not very thick.
21. Because unsectioned samples are thick, mounting medium that stays liquid is not an option. Hardening mounting medium helps the tissue and coverslip stay in place. Preferably use a mounting medium that prevents photobleaching and preserves signal for extended storage. ProLong Antifade Mountant (Thermo Fischer Scientific, Waltham, MA) is an example of a mounting medium that works very well.
22. Hardening mounting medium tends to contract as it dries. For this reason, it is important to add abundant amounts of mounting medium or, once dry, air will advance into the sample, forming empty spaces in the tissue that will not be possible to image. Similarly, small bubbles left when mounting will expand as the mounting medium dries, making image acquisition difficult.
23. To a great extent, the quality of the image will depend on intrinsic properties of the objective, such as numerical aperture (NA, typically specified as a dimensionless number on the objective itself). The higher the NA, the more resolving power of an objective. Regarding magnification per se, always start with lower magnification, but be aware that unless the staining is extremely bright, no signal will be detected when

using low-power objectives (e.g. 2× or 4×). Magnification using a 20× objective provides a good balance between obtaining good overall tissue context and good resolution at a cellular level. A 20× objective should allow for visualization of host cells and *Y. pestis* as well, provided the latter is bright enough. To visualize *Y. pestis* in more detail, a 60× objective might be necessary.

24. While, ultimately, images will be acquired under the excitation of a laser, microscopes are equipped with additional excitation (i.e. mercury or LED excitation sources) to allow for direct observation of samples through the eyepiece of the microscope, as opposed to through a computer screen. It is easier to observe a sample directly through the eyepiece than through a screen and, thus, it is recommended to use this method to survey the tissue in *X*, *Y*, and *Z*, testing different magnifications.

25. As previously stated, LSM allows for acquisition in the form of a single optical slice, or multiple consecutive optical slices (all with same *X* and *Y* coordinates, but different values in *Z*), forming a stack. The advantages of a stack over a single slice are: (a) more opportunities to choose the best slice for a final image, (b) with the appropriate software, a stack can be represented as a three-dimensional image or animation, allowing for more complex manipulations, such as rotation, etc.

26. Once an area is selected for imaging (*X* and *Y* coordinates), two *Z* values have to be specified in the operating software of the microscope; one value corresponds to the position of the first optical slice and the other value corresponds to the last. The thickness of a stack will be defined by the distance between these two optical slices.

27. Penetrance will depend on the technology being used. As previously discussed, two-photon LSM has more penetrance power than confocal LSM. In addition, wavelengths closer to red penetrate more efficiently through tissues than wavelengths closer to blue. For this reason, fluorophores that are excited at wavelengths closer to red are easier to visualize at a specific depth in comparison to fluorophores excited at wavelengths closer to blue. Lastly, some microscopes allow for preset increments in laser excitation as acquisition progresses deeper into a tissue. This can compensate for the loss of signal due to depth, but phototoxicity can be a problem if laser intensities are too high.

28. When defining step size, consider the level of resolution in the *Z* dimension needed for a proper observation. If a phagocyte is to be imaged, consider these cells can be around 15–20 μm in diameter. Thus, if the step size is 10 μm, a slice will cut through

a cell only once. This might be enough to report the cell is present, but not enough to visualize it in more detail, or to generate a three-dimensional image of the entire cell. Moreover, *Y. pestis* measures about 2 μm. Step sizes of 10 μm will be too large and will skip a significant number of bacteria. On the other hand, too small of a step size might result in imaging distortion. A recommended starting point is step sizes between 1 μm and 5 μm.

29. Longer dwell times result in more light exciting a fluorophore, and consequently more photons being collected per pixel, all of which contributes to a good quality image. However, longer dwell times can increase phototoxicity and photobleaching.

30. As with dwell times, higher laser power will result in more light exciting a fluorophore, contributing to a better quality image. Consequently, higher laser power will also increase phototoxicity and photobleaching. PMTs are exceptionally sensitive detectors that detect extremely low levels of light. Because PMTs work based on the photons they receive from an excited fluorophore, changes in the gain settings do not affect the sample. That is, no photobleaching or phototoxicity occurs. Higher PMT gain settings translate into greater amplification of signal and more sensitivity, but they also result in higher noise signal (not to be mistaken with background signal, which comes from the sample itself, not the detectors).

31. Fiji [12] and Imaris (Belfast, UK) are very good options for the visualization of microscopy datasets and for their analysis. Fiji offers a very large number of functions for measurements and other types of analysis, and its open source platform makes it a very popular option. Imaris is relatively expensive, but many microscopy cores offer a license for authorized users. While Fiji provides excellent solutions for an endless number of needs, Imaris dominates when it comes to the visualization and rendering of three-dimensional images from optical stacks.

32. In some cases, a color has to be assigned to each channel manually. It is important to remember that LSM images are not acquired in the context of true color technology. Each channel is acquired in grayscale. Thus, a color must be assigned digitally to differentiate each channel, especially once all channels are merged into the same image. Consequently, regardless of the color that is chosen for each channel, the image will depict false color.

33. While it is rare, analysis software might not display the correct numbers in the scale bar. An easy way to determine the scale bar shows an accurate measurement is to use common knowledge about the size of cells. For instance, a dendritic cell should measure around 20 μm and the scale bar should reflect this measurement. A value of 60 μm for a dendritic cell indicates the

*X* and *Y* values of the image (size) need to be corrected by a factor of three.

34. Software for visualization and analysis of microscopy datasets give many options for the presentation of a final image. A single optical slice is the simplest form to present an image. A whole optical stack can be presented in three basic forms: The three-dimensional stack can be depicted as a single two-dimensional image in what is called a maximum intensity projection or MIP. A MIP will display the voxels (three-dimensional units analogous of pixels) of maximum intensity in comparison with other voxels in the same projection line, from the viewpoint to the projection plane. A stack can also be represented as a two dimensional image showing a specific position of the original three-dimensional stack. Lastly, the stack can also be represented in three dimensions, in movement, and as an animation. These animations typically show the sample rotating or moving from side to side to make depth more evident. It is a common mistake to represent stacks rotating at very high speeds.

## Acknowledgments

The author would like to thank Virginia L. Miller for kindly reviewing and editing this document.

### References

1. Fischer RS, Wu Y, Kanchanawong P, Shroff H, Waterman CM (2011) Microscopy in 3D: a biologist's toolbox. Trends Cell Biol 21:682–691
2. Gonzalez RJ, Lane MC, Wagner NJ, Weening EH, Miller VL (2015) Dissemination of a highly virulent pathogen: tracking the early events that define infection. PLoS Pathog 11: e1004587
3. Shannon JG, Bosio CF, Hinnebusch BJ (2015) Dermal neutrophil, macrophage and dendritic cell responses to *Yersinia pestis* transmitted by fleas. PLoS Pathog 11:e1004734
4. Shannon JG, Hasenkrug AM, Dorward DW, Nair V, Carmody AB, Hinnebusch BJ (2013) *Yersinia pestis* subverts the dermal neutrophil response in a mouse model of bubonic plague. MBio 4:e00170–e00113
5. Helmchen F, Denk W (2005) Deep tissue two-photon microscopy. Nat Methods 2:932–940
6. Miyawaki A, Shcherbakova DM, Verkhusha VV (2012) Red fluorescent proteins: chromophore formation and cellular applications. Curr Opin Struct Biol 22:679–688
7. Richardson DS, Lichtman JW (2015) Clarifying tissue clearing. Cell 162:246–257
8. Ariel P (2017) A beginner's guide to tissue clearing. Int J Biochem Cell Biol 84:35–39
9. Azaripour A, Lagerweij T, Scharfbillig C, Jadczak AE, Willershausen B, Van Noorden CJF (2016) A survey of clearing techniques for 3D imaging of tissues with special reference to connective tissue. Prog Histochem Cytochem 51:9–23
10. Shaner NC, Steinbach PA, Tsien RY (2005) A guide to choosing fluorescent proteins. Nat Methods 2:905–909
11. Hadjantonakis A-K, Dickinson ME, Fraser SE, Papaioannou VE (2003) Technicolour transgenics: imaging tools for functional genomics in the mouse. Nat Rev Genet 4:613–625
12. Schindelin J, Arganda-Carreras I, Frise E, Kaynig V, Longair M, Pietzsch T, Preibisch S, Rueden C, Saalfeld S, Schmid B, Tinevez J-Y, White DJ, Hartenstein V, Eliceiri K, Tomancak P, Cardona A (2012) Fiji: an open-source platform for biological-image analysis. Nat Methods 9:676–682

# Chapter 7

# Real-Time Monitoring of *Yersinia pestis* Promoter Activity by Bioluminescence Imaging

**Anne Derbise, Olivier Dussurget, Elisabeth Carniel, and Javier Pizarro-Cerdá**

## Abstract

Bioluminescence imaging (BLI) has become a major strategy for real-time analysis of dynamic biological processes. In particular, bioluminescent reporter microorganisms have been designed to advance our understanding of infectious diseases. Non-invasive monitoring of light-emitting pathogenic bacteria has revealed novel features of pathogenesis and enabled quantitative and qualitative analysis of antibacterial therapies. Transcriptional gene fusions using the bacterial luciferase operon *luxCDABE* as a reporter have been successfully used to monitor gene expression in vitro and in vivo, leading to valuable applications and major findings. In this chapter, we describe the construction of *Yersinia pestis* strains bearing a chromosomal copy of the *luxCDABE* operon under the control of promoters regulated by temperature and their application to quantify gene expression in real-time in bacteria growing in vitro and in a murine bubonic plague model.

**Key words** Bioluminescence, Real-time imaging, Reporter system, Promoter activity, Plague, *Yersinia pestis*, Mice

## 1 Introduction

Bioluminescence is the production and emission of light by a living organism. With the development of highly sensitive photon detectors, in vivo bioluminescence imaging (BLI) techniques have been particularly useful for real-time noninvasive tracking of bacteria within animal models.

Bioluminescent imaging has allowed monitoring of the proliferation and dissemination of *Yersinia pestis* in real-time in different murine models, such as bubonic and pneumonic plague [1–3]. In these studies, bioluminescence was produced by the *Photorhabdus luminescens* luciferase which is encoded by the genes *luxA* and *luxB* that catalyzes the oxidation of a reduced flavin mononucleotide by oxygen leading to the production of photons that can be quantified. The substrate is recycled by the enzymatic complex encoded

Viveka Vadyvaloo and Matthew B. Lawrenz (eds.), *Pathogenic Yersinia: Methods and Protocols*, Methods in Molecular Biology, vol. 2010, https://doi.org/10.1007/978-1-4939-9541-7_7, © Springer Science+Business Media, LLC, part of Springer Nature 2019

by genes *luxD, luxC,* and *luxE,* allowing detection of bacteria expressing the *luxCDABE* operon without the addition of exogenous substrate. The *P. luminescens luxCDABE* operon provides several advantages when expressed in *Y. pestis.* It is compatible with the in vivo mouse body temperature and does not impact fitness of *Y. pestis.* The turnover production and activity of the luciferase A and B subunits allow in vivo multiplication monitoring and correlate with the bacterial load when a constitutive promoter is fused with the *lux* operon. The short half-life of *P. luminescens* luciferase prevents signal accumulation and makes it appropriate for quantitative studies.

In this chapter, we describe the use of bioluminescence for real-time in vitro and in vivo monitoring of *Y. pestis* promoter activity. To allow the comparative study of gene expression and to avoid side effects from plasmid variable copy numbers and plasmid loss in vivo, we constructed a single chromosomal version of the *P. luminescens luxCDABE* operon fused downstream of the *Y. pestis* promoter regions of interest. The mini Tn*7* system was used for specific chromosomal insertion at the *att*Tn*7* site between two housekeeping genes, *glmS* and *pstS,* encoding glucosamine-6-phosphate synthetase and periplasmic phosphate-binding protein, respectively [4]. Transposition of mini Tn*7* in this intergenic region does not modify the virulence of *Y. pestis* [2].

Here, we validate and evaluate the sensitivity of the bioluminescent reporter system using the promoters of the *caf1M* and *YPO0499* genes, which are known to be regulated by temperature. The *caf1M* gene is part of the *Y. pestis* F1 capsule encoding operon whose expression is up-regulated at 37 °C, while the gene *YPO0499* is part of operon encoding the type VI secretion system whose expression is down-regulated at 37 °C. The constitutively active promoter of *rplN,* encoding a protein of the 50S ribosomal subunit was used as a control.

## 2 Materials

### 2.1 Construction of Y. pestis Reporter Strains

1. Fully virulent *Y. pestis* CO92 strain isolated from a fatal human pneumonic plague case [5] (*see* **Note 1**).
2. *Escherichia coli* BW19610 strain [6].
3. pUC18R6KT mini-Tn*7*-*Km*$^r$-*luxCDABE* (Fig. 1) a non-replicative plasmid in *Yersinia* (*see* **Note 2**), which requires a *pir* expressing strain, such as *Escherichia coli* BW19610, for its replication.
4. Transposase-encoding plasmid pTNS2 [4] (Fig. 1a) (*see* **Note 3**).
5. Flp recombinase-encoding plasmid pFLP3 [4] (Fig. 1b).

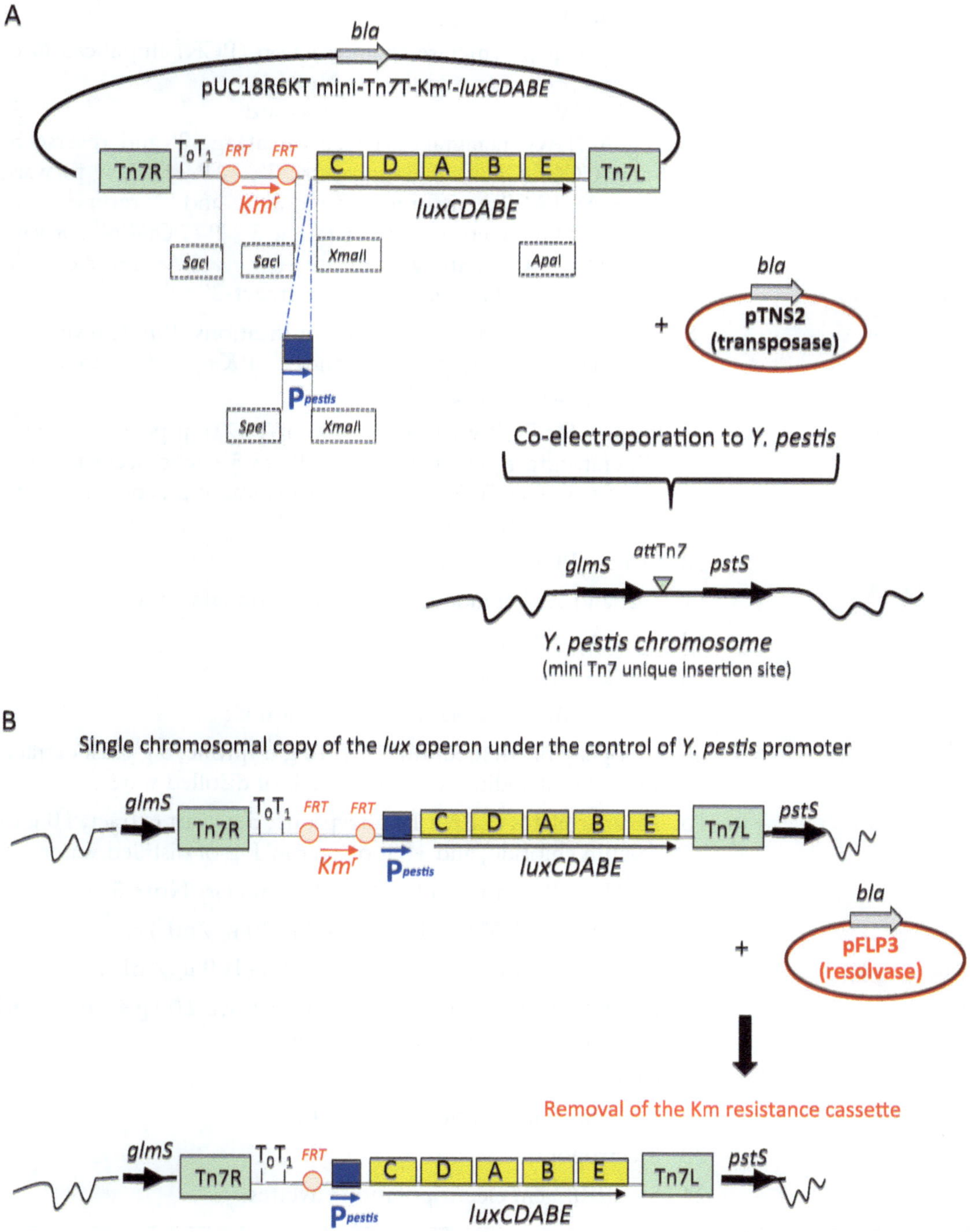


**Fig. 1** Construction of the bioluminescent reporter vector to study *Y. pestis* promoter activity. (**a**) Suicide delivery pUCR6K-mini Tn*7* based vector. Tn*7*L and Tn*7*R, left and right ends of Tn*7*; $T_0T_1$, transcriptional terminators; *bla*, β-lactamase-encoding gene; $Km^R$ kanamycin resistance encoding cassette flanked by FRT, Flp recombinase target; $P^{pestis}$, *Y. pestis* promoter regions; *luxCDABE*, *Photorabdus luminescens lux* operon where *luxA* and *luxB* encode the subunits A and B of bacterial luciferase, while *luxC*, *luxD*, and *luxE* encode the subunits of a fatty acid reductase that provides the fatty aldehydes to the luciferase. Only enzyme restriction sites used for cloning $Km^R$, promoter regions and *lux* operon are indicated. pTNS2, helper suicide plasmid

6. Thermal Cycler.
7. Primers for polymerase chain reaction (PCR) amplification of *Y. pestis* promoter regions (*see* **Note 4**).
   P*rplN*, forward 5-′-cgcACTAGTtagagtagccaactcgaacataataagc-3′ and reverse 5-′-ataCCCGGGgctccgctaatgtcaactac-3′; P*caf1M*, forward 5′-cgcACTAGTcttattttggctattcatcc-3′ and reverse 5-′-tgaCCCGGGgcttgctattccttgctattcg-3′; P*YPO0499*, forward 5′-cgcACTAGTcttattttggctattcatcc-3′ and reverse 5-′-tgaCCCGGGGcgttgctattccttgctattcgct-3′.
8. Primers for promoter sequence verifications: P*luxC* reverse 5-′-cacgaatgtatgtcctgcgtc-3′ and P*Km* forward 5-′-tcctatggaactgcctcggt-3′.
9. Primers for PCR verification of mini-Tn*7* transposition: P*glmS* 5′-gctatacgtgtttgctgatcaagatg-3′, P*pstS* 5′- acgccaccggaagaacc-gatacct-3′, PTn*7*L 5′-attagcttacgacgctacaccc-3′, and PTn*7*R 5′-cacagcataactggactgatttc-3′.
10. Qiagen PCR purification kit.
11. *Spe*I and *Xma*I endonuclease restriction enzymes.
12. T4 DNA ligase.
13. Sterile distilled water.
14. Qiagen Midiprep plasmid purification kit.
15. Lysogeny broth medium (LB): 10 g tryptone, 5 g yeast extract, and 10 g of sodium chloride in 1 L of distilled water.
16. LB agar plates (LBA): 10 g tryptone, 5 g yeast extract, 10 g of sodium chloride, and 15 g of agar in 1 L of distilled water.
17. LBAH: LBA with 0.002% (w/v) hemin (*see* **Note 5**).
18. LBAHKm: LBAH with kanamycin (30 μg/mL).
19. LBAHCarb: LBAH with carbenicillin (100 μg/mL).
20. LBAHSuc: 10 g tryptone, 5 g yeast extract, 100 g sucrose, and 15 g of agar in 1 L of distilled water.
21. Cold, sterile 10% glycerol.
22. 1 × Phosphate-buffered saline (PBS).
23. Electroporator.
24. Sterile 0.2 cm electroporation cuvettes.

**Fig. 1** (continued) encoding the specific TnsABC+D transposition pathway. Transposition of mini Tn*7*-Km$^R$-P$^{pestis}$-*lux* into *Y. pestis* chromosome is achieved by co-electroporation of competent *Y. pestis* CO92 cells with pTNS2 and pUCR6K-KmR-*lux*. The *att*Tn*7* unique insertion site located in the 482-bp *glmS*- *pstS* intergenic region. (**b**) Removal of the Km$^R$ cassette. Mini Tn7-Km$^R$-P$^{pestis}$-*lux* reporter sytem integrated as a monocopy in *Y. pestis* chromosome; pFLP3, Flp recombinase encoding plasmid. Plasmid pFLP3 is electroporated into the bioluminescent *Y. pestis* CO92 recombinant strain to remove the KmR cassette in order to avoid any transcriptional interference with the bioluminescence reporter system

### 2.2 In Vitro Measurement of Promoter Activity

1. LB.
2. PBS.
3. 250 mL Erlenmeyer flask.
4. Incubators set to 28 °C and 37 °C.
5. Spectrophotometer and cuvettes.
6. Plate reader capable for photon emission detection at 490 nm.
7. 96-well opaque white plate.

### 2.3 Measurement of Promoter Activity in a Murine Bubonic Plague Model

1. 7-week-old female OF1 mice.
2. Sterile 26 G × 1/2″ needles and 1 mL syringes.
3. Sterile surgical dissection instruments.
4. Tissue and cell disruption homogenizer (Tissuelyser II, Qiagen).
5. Sterile 2 mL polypropylene microtube with cap filled with 6–8 glass beads (3 mm diameter, Omni international) for disruption of lymph nodes or 2 metal beads (2.4 mm, Omni international) for skin disruption.
6. $CO_2$.
7. Anesthesic mix in PBS: ketamine 20 mg/mL (Imalgene 1000, Bayer) and xylazine 2 mg/mL (Rompun 2%, Merial).
8. Poly(methyl methacrylate) confinement box (TEM SEGA) (*see* **Note 6**).
9. Imaging System (IVIS Spectrum, Perkin Elmer).
10. Living Image 4.3.1 software (Caliper Life Sciences).

## 3 Methods

### 3.1 Promoter Amplification and Cloning to Generate a Bioluminescent Reporter Plasmid (Fig. 1a)

1. Amplify the promoter region of interest using standard PCR methods with forward and reverse primers for $P^{rplN}$, $P^{caf1M}$, and $P^{YPO0499}$ regions.
2. Purify PCR products using the Qiagen PCR purification kit.
3. Digest the purified PCR product and the pUC18R6KT mini-Tn*7-Km*$^r$*-luxCDABE*-based vector with *Spe*I and *Xma*I following the enzyme manufacturer's instructions.
4. Ligate the digested PCR product and plasmid following the manufacturer's instructions with T4 DNA ligase overnight at 16 °C.
5. Electroporate 1–10 μL of the ligation reaction into *Escherichia coli* BW19610 electrocompetent cells [7]. Add 1 mL of LB and incubate at 37 °C for 45 min. Plate various bacterial

suspensions (pure to 1/100 dilution) on LBA supplemented with 100 μg/mL carbenicillin and incubate overnight at 37 °C.

6. Check the correct insertion and sequence of the promoter of interest by restriction enzyme digestion using *Spe*I and *Xma*I and by sequencing using primers P*Km* and P*luxC*. The recombinant plasmid will be referred to in this chapter as pUC18-mini-Tn*7*-K-P$^{pestis}$-*lux*.

### 3.2 Construction of a Single-Copy Bioluminescent Reporter Gene Expression System in Y. pestis (Fig. 1)

1. Purify plasmids pTNS2 (transposase provider) and pUC18-mini-Tn*7*-K-P$^{pestis}$-*lux* using the Qiagen Midiprep plasmid purification kit.
2. Prepare electrocompetent *Y. pestis* CO92 cells resuspended in ice-cold 10% glycerol (*see* **Note 7**).
3. Electroporate 50 μL of electrocompetent *Y. pestis* cells with 400 ng of each plasmid using the following settings: 2.5 kV, 200 Ω, and 25 μF.
4. Add 1 mL of LB and incubate at 28 °C under agitation for 90 min.
5. Plate various dilutions on LBAHKm and incubate for 48 h at 28 °C (recommended dilutions, 1/10–1/10$^3$).
6. Verify the correct insertion of mini-Tn*7*-K-P$^{pestis}$-*lux* (Fig. 1) into the chromosomal *glmS-pstS* intergenic region by PCR using the pairs of primers P*glmS*/PTn*7*R, P*pstS*/PTn*7*L, and P*glmS*/ P*pstS* flanking the *att*Tn*7* insertion site (*see* **Note 8**).
7. Prepare *Y. pestis* CO92:: mini-Tn*7*-K-P$^{pestis}$-*lux* electrocompetent cells resuspended in ice-cold 10% glycerol (*see* **Note 7**).
8. Remove the kanamycin resistant cassette by electroporation of 50 μL electrocompetent *Y. pestis* CO92:: mini-Tn*7*-K-P$^{pestis}$-*lux* cells with 100 ng of plasmid pFLP3 (Fig. 1b).
9. Select the recombinant pFLP3 harboring cells by plating various dilutions on LBAHCarb and incubate 48 h at 28 °C.
10. Replicate plate individual colonies on both LBAHKm and LBAH plates to identify the clones that have lost the kanamycin resistance gene.
11. Eliminate the pFLP3 plasmid by growing *Y. pestis* CO92:: mini-Tn*7*-P$^{pestis}$-*lux* on LB overnight at 28 °C.
12. Serial dilute the overnight culture and plate 100 μL of 1/100–1/10$^6$ dilutions on LBAHSuc plates.
13. Replicate plate individual colonies on both LBAHCarb and LBAH to identify the clones that have lost pFLP3.
14. Check for the presence of known unstable *Y. pestis* virulence genetic elements such as plasmids, pYV, and pPla and the high pathogenicity island by PCR [8] (*see* **Note 9**).

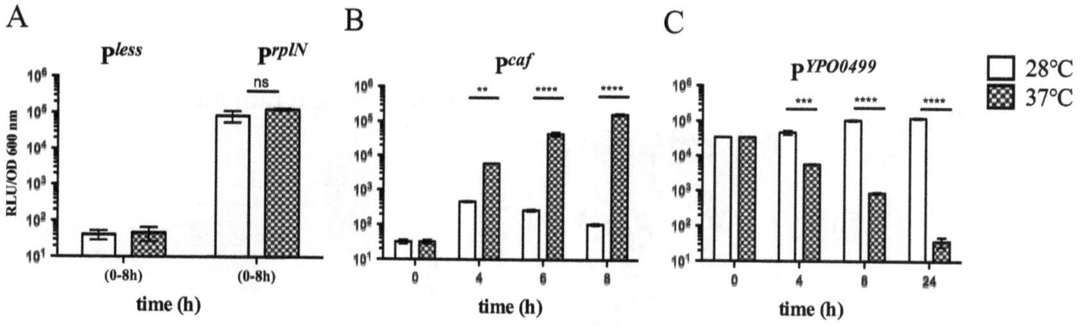


**Fig. 2** Detection of bioluminescence in vitro. *Yersinia pestis* C092 recombinant strains were pre-grown on LBAH plate. Then bacterial cells were resuspended in PBS, optical density at 600 nm was adjusted and a 50 mL LB flask was inoculated 1/100e to obtain approximately 10e7 CFU/mL and incubated under agitation. Optical density and photon emission were recorded over time by taking a 1 mL and 0.1 mL aliquot of the growth suspension, respectively. (**a**) Detection of bioluminescence signals obtained with strains C02:: Tn*7*-$P^{less}$-*lux* and Tn*7*-$P^{rplN}$-*lux* when grown in LB at 28 °C and 37 °C. The C092:: Tn*7*-$P^{less}$-*lux* strain was used as a negative control to evaluate in each growth condition the bioluminescence background. The promoter region of the *rplN* ribosomal encoding gene was used as a constitutively active promoter in each growth condition. Since the ratio RLU over CFU did not vary significantly over time or with temperature for the two control strains, only an average of the signals measured in each growth condition is indicated. Detection of *Y. pestis* $P^{caf}$ (**b**) and $P^{YPO0499}$ (**c**) activities, two temperature-sensitive promoters in LB at 28 °C and 37 °C. The increase of bioluminescence reflects the activity of the $P^{caf}$ promoter showing up-regulated *Y. pestis* capsule expression, while the decrease of bioluminescence reflects the down-regulated expression of the type VI secretion *YPO0499* gene at 37 °C. Statistical analyses were performed with an unpaired *t*-test ($^{****}p < 0.0001$, $^{***}p < 0.0003$, and $^{**}p < 0.001$) using Prism 5 GraphPad software

### 3.3 In Vitro Detection of Promoter Activity

1. Make a lawn of *Y. pestis* recombinant strain (CO92:: mini-Tn*7*-K-$P^{pestis}$-*lux*) by spreading an aliquot from a −80 °C frozen stock and incubating overnight at 28 °C on LBAH.
2. Resuspend bacterial cells in fresh 10 mL LB by scraping from the bacterial lawn, adjust optical density to 1, dilute 1/100 in 50 mL LB and grow under constant agitation (*see* **Note 10**). In this study, two temperatures were tested, 28 °C and 37 °C (Fig. 2).
3. Measure the optical density at 600 nm over time using 1 mL samples (Fig. 2).
4. Measure the photon emission/s (RLU) over time by transferring 0.1 mL sample from the 50 mL growth in a 96 well plate (Fig. 2) (*see* **Note 11**).
5. To determine promoter activity, divide the number of photon/s (RLU) by the absorbance at 600 nm in exponential phase to correct for the relative differences in cell number between strains.

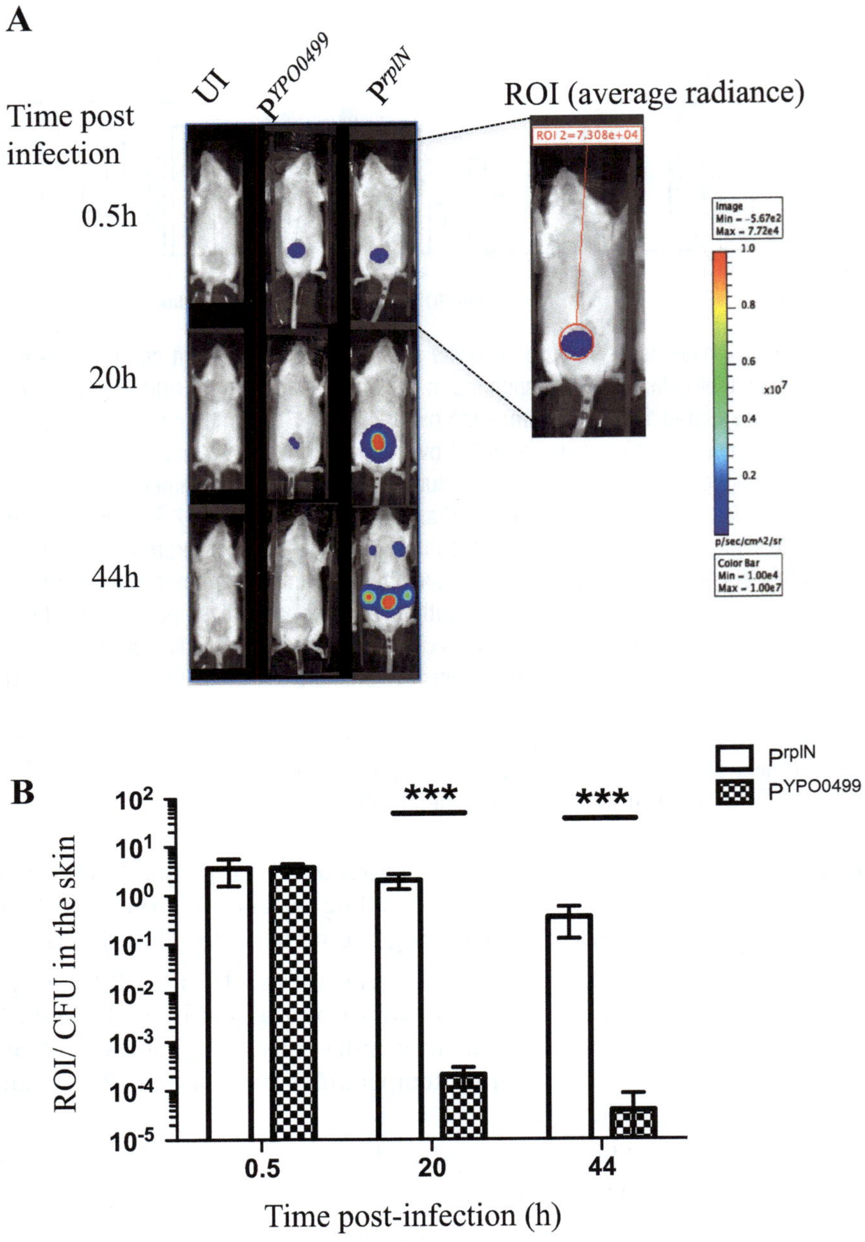


**Fig. 3** Bioluminescence imaging of *Y. pestis*-infected mice. (**a**) OF1 mice were injected subcutaneously in the linea alba with 10e5 CFU of *Y. pestis* CO92:: Tn*7*-P$^{rplN}$-*lux* or P$^{YPO0499}$-*lux* and imaged at different time points using an IVIS Spectrum imaging system. Sequential images from representative animals are shown. The color scales represent the light emission and is kept constant throughout the observation period. In this study, a small binning and an acquisition time of 10–240 s were used. Uninfected mice (UI) were used to set the light emission background. Regions of interest (ROI) were drawn using the Living Image 4.3.1 software and average bioluminescence (photon/s/cm$^2$/sr) was calculated for the injection site, as shown here for the P$^{rplN}$-fusion at 0.5 h post injection in the skin. Bioluminescence signal is detected in the skin at the site of injection. The increased signal observed with the P$^{rplN}$-*lux* fusion construct reflects *Y. pestis* multiplication. In addition, bacterial dissemination to the inguinal and axillary draining lymph nodes is detected 44 h postinfection (**b**)

### *3.4 In Vivo Detection of Y. pestis Promoter Activity in a Murine Bubonic Plague Model*

Animal experiments must receive approval from national and institutional committees.

Our protocol was approved by the French committee CETEA (n°2014–0025) and Institut Pasteur CHSCT (n°0399).

1. Grow bioluminescent *Y. pestis* on LBAH plates overnight at 28 °C, as described in Subheading 3.3, **step 1**.
2. Resuspend bacteria in PBS and adjust OD600 to 0.1, as described in Subheading 3.3, **step 2**.
3. Prepare serial dilutions in PBS.
4. Inject the required amounts of bacterial cells within a maximal volume of 50 μL for subcutaneous injection (*see* **Note 12**). In the experiment shown in Fig. 3, 10e5 colony forming units (CFU) were injected (*see* **Note 13**).
5. At desired time points, anesthetize the mice by injecting intraperitoneally 100 μL of the anesthetic mix (ketamine/xylazine).
6. Place mice in a confinement box (*see* **Note 6**).
7. Place the confinement box into the IVIS imaging chamber with the mice injection site facing the CCD camera.
8. Choose the field of view according to the number of animals to be imaged (*see* **Note 14**).
9. Set up the imaging acquisition conditions on the control panel by selecting luminescent imaging mode, block excitation filter, and open emission filter.
10. Adjust the binning to "small" and the exposure time according to the intensity of the signal (between 5 s and 2 min exposure time) (*see* **Note 15**).
11. Click "acquire" to measure luminescence (Fig. 3).
12. Use the tool palette/image adjust to adjust the color scale in order to visualize all sites that exhibit signals above the background (*see* **Note 16**).
13. Create regions of interest (ROI) by using the ROI tool (*see* **Note 17**) and measure ROIs.
14. In order to evaluate the background bioluminescent signal, measure the number of photons emitted from the same ROI

**Fig. 3** (continued) Evaluation of promoter activity during *Y. pestis* multiplication in the skin of OF1 mice. Mice were sacrificed at indicated time points. The site of injection was dissected and the skin was homogenized to determine bacterial loads (CFU) and compare them to the measured ROI (photon/s/cm$^2$/sr). The P$^{rplN}$-*lux* fusion illustrates the constitutive expression of *rplN*, where the ROI/CFU ratio is similar over time. The P$^{YPO0499}$-*lux* fusion demonstrates the decrease of *YPO0499* expression when the pathogen multiplies in the skin as exemplified by the decrease ROI/CFU ratio of a non-infected mouse placed in the same confinement box as infected mice (*see* **Note 16**).

15. Calculate the luminescence to determine changes in luciferase activity over time.

    Promoter activity is measured by counting the photons emitted by the bacteria. To allow comparisons between signals express the ROI bioluminescence in average radiance (photons/s/$cm^2$/steradian).

16. To determine the number of bacteria in the infected tissues, euthanize mice with $CO_2$.
17. Harvest the tissues aseptically and place them in a 0.5 mL sterile PBS tubes containing glass or metal beads.
18. Apply a 2 min tissue disruption cycle of 30 agitations/s with the TissueLyser II Qiagen apparatus (*see* **Note 18**).
19. Spread 50 μL of serial dilutions of the cell suspension on LBAH plates and incubate 48 h at 28 °C for colony counts.
20. Results are given in ROI per CFU (Fig. 3b).

## 4 Notes

1. Experiments performed on fully virulent *Y. pestis* require a Biosafety Level 3 laboratory and animal facility.
2. The plasmid pUC18R6KT mini-Tn*7*-*Km*$^r$-*luxCDABE* was constructed by subcloning (1) the Km resistance cassette from (Km$^r$) pFKM1 [4] into the *Sac*I restriction site and (2) the *P. luminescens luxCDABE* operon into the *Apa*I/*Xma*I restriction sites of pUC18R6KT mini-Tn*7* [4]. The resulting plasmid does not contain any promoter region for *luxCDABE* expression, and is used for evaluation of bioluminescence background in in vitro studies, as well as for subsequent cloning of the different *Y. pestis* promoters to be studied.
3. Plasmid pTNS2 is not replicative in *Y. pestis*. Efficient production of the transposase immediately after pTNS2 introduction in *Y. pestis* allows transposition of the mini Tn*7*. Loss of the transposase encoding plasmid will prevent subsequent transposition of Tn*7*. Verification of the loss of pTNS2 after Tn*7* transposition in *Y. pestis* chromosome is performed by testing carbenicillin susceptibility on LBAHCarb.
4. For a directional cloning of promoter regions upstream of the *luxCDABE* operon, forward and reverse primers contain 5′ *Spe*I and *Xma*I restriction sites, respectively (Fig. 1a).
5. When colony forming units (CFU) count have to be performed on *Y. pestis* suspension, it is recommended to supplement the agar lysogeny broth medium with 0.002% (w/v) porcine hemin. The presence of hemin in the medium will allow

individual bacteria to grow and form a CFU within 48 h at 28 °C. If hemin is omitted colonies will be small, difficult to count and the number of colonies could be underestimated.

6. When animals are infected with a class 3 pathogen, their transportation and visualization in the IVIS Spectrum System should be conducted in a biosafety containment box. In our study, we used a poly(methyl methacrylate) confinement box (TEM SEGA) to warrant the confinement of infected animals and to allow the photon acquisition by the CCD detector.
7. To prepare electrocompetent *Y. pestis*, bacteria were grown overnight at 28 °C in LB. The culture was diluted (1/50) in fresh broth and grown until reaching A600 = 0.8. Bacterial cells were harvested by centrifugation (4000 × *g*, 15 min, and 4 °C), washed twice in cold sterile distilled water and once in 10% glycerol. The pellet was resuspended in 10% glycerol to reach a concentration factor of 500 times compared to the original growth volume. Electrocompetent cells were aliquoted in 50 μL and frozen at −80 °C until use.
8. Correct insertion of the mini-Tn7 transposon is verified by the amplification of 770 bp and 336 bp PCR fragments with primers P*pstS*/PTn*7*L and P*glmS*/P*pstS*, respectively and absence of amplification with primers *glmS-pstS* when the following standard PCR program is used: 1 cycle at 95 °C for 3 min, 35 cycles (1) at 95 °C for 30 s, (2) at 55 °C for 30 s, (3) at 72 °C for 1 min, and finally 1 cycle at 72 °C for 3 min.
9. Presence of the pYV and pPla plasmids and the high pathogenicity island are verified by PCR amplifications of 565 bp *yopM*, 480 bp *pla*, and 434 bp *irp2* fragments using primer pairs YopM-f (5′-ataactcatcgggggcaaaat-3′)/YopM-r (5-′-gcgttatttatccgaatttagc-3′), pla-f (5′-atcttactttccgtgagaag-3′)/ pla-r (5′-cttggatgttgagcttccta-3′), and irp2-f (5-′-tgatgagatggttacactgg-3′)/irp2-r (5′-gtcgttatggcgtaaatcat-3′), respectively. The standard PCR program indicated in **Note 8** is used.
10. Conditions of growth culture may vary depending on the known or putative activity of the promoter region studied.
11. Since *Y. pestis* tends to aggregate when grown in LB, we recommend to perform the in vitro growth in a large volume (50 mL LB) in a 250 mL Erlenmeyer flask under 140 rpm agitation and measure the absorbance overtime by taking 0.1 mL samples.
12. To facilitate the subcutaneous injection, the fur of the animal can be locally shaved 24 h before the injection. Depending on the site of injection anesthesia can be required prior to the injection. In the experiment shown in Fig. 3, the animals were not anesthetized.

13. The volume and dose injected can vary depending on the injection mode, site and the expected level of signal at early time points of the infection.
14. IVIS spectrum allows the use of different fields of view. When more than three animals are imaged use field of view D. Sensitivity of the detection camera may vary depending on the field of view chosen during acquisition. If a comparative study is performed make sure to use the same field of view throughout the experiment.
15. For quantification, make sure to use an exposure time that does not lead to signal saturation. In the image display, a message indicates saturated signals. In most of our experiments, the exposure time varied from 5 to 240 s with a small binning. If a weak signal is expected, the exposure time can be increased up to 5 min with a large binning.
16. When several infected animals are to be imaged within the same containment box, it is important to evaluate the level of the background signal by imaging a non-infected animal hosted in the same box. This level may change depending on the intensity of the overall signal within the same box.
17. The shape and size of the ROI will be adjusted to the shape and size of the site exhibiting a bioluminescent signal. In the case of the site of injection shown in Fig. 3, the shape and size of the ROI were drawn by using the function "duplicate."
18. For tissue disruption, other homogenizer apparatuses can be used. If so, validation of the efficiency of bacterial count must be performed.

## Acknowledgments

We thank Marie-Anne Nicola for her advice in the use of the bioluminescence system at the Photonic BioImaging facility of the Institut Pasteur.

### References

1. Nham T et al (2012) Imaging of bubonic plague dynamics by in vivo tracking of bioluminescent *Yersinia pestis*. PLoS One 7(4):e34714
2. Sun Y et al (2012) Development of bioluminescent bioreporters for in vitro and in vivo tracking of *Yersinia pestis*. PLoS One 7(10):e47123
3. Gonzalez RJ et al (2012) Bioluminescence imaging to track bacterial dissemination of *Yersinia pestis* using different routes of infection in mice. BMC Microbiol 12(147)
4. Choi KH et al (2005) A Tn*7*-based broad-range bacterial cloning and expression system. Nat Methods 2(6):443–448
5. Doll JM et al (1994) Cat-transmitted fatal pneumonic plague in a person who traveled from Colorado to Arizona. Am J Trop Med Hyg 51 (1):109–114
6. Metcalf WW, Jiang W, Wanner BL (1994) Use of the rep technique for allele replacement to construct new *Escherichia coli* hosts for maintenance

of R6K gamma origin plasmids at different copy numbers. Gene 138(1–2):1–7

7. Dower WJ, Miller JF, Ragsdale CW (1988) High efficiency transformation of *E. coli* by high voltage electroporation. Nucleic Acids Res 16(13):6127–6145

8. Tsukano H et al (1996) Detection and identification of *Yersinia pestis* by polymerase chain reaction (PCR) using multiplex primers. Microbiol Immunol 40(10):773–775

# Chapter 8

# Discovering *Yersinia*–Host Interactions by Tissue Dual RNA-Seq

Maria Kusmierek, Ann Kathrin Heroven, Michael Beckstette, Aaron M. Nuss, and Petra Dersch

## Abstract

A detailed knowledge about virulence-relevant genes, as well as where and when they are expressed during the course of an infection is required to obtain a comprehensive understanding of the complex host–pathogen interactions. The development of unbiased probe-independent RNA sequencing (RNA-seq) approaches has dramatically changed transcriptomics. It allows simultaneous monitoring of genome-wide, infection-linked transcriptional alterations of the host tissue and colonizing pathogens. Here, we provide a detailed protocol for the preparation and analysis of lymphatic tissue infected with the mainly extracellularly growing pathogen *Yersinia pseudotuberculosis*. This method can be used as a powerful tool for the discovery of *Yersinia*-induced host responses, colonization and persistence strategies of the pathogen, and underlying regulatory processes. Furthermore, we describe computational methods with which we analyzed obtained datasets.

**Key words** Tissue dual RNA-seq, Host–pathogen interaction, *Yersinia* infection, Host-adapted metabolism, Non-coding RNAs, Transcriptomics

## 1 Introduction

The interplay between enteric pathogens, such as the enteropathogenic *Yersinia* species *Y. enterocolitica* and *Y. pseudotuberculosis*, and their host is a very complex and highly dynamic process. Both interacting organisms initiate intricate programs to adapt rapidly to the new situation and ensure their survival. For instance, enteric yersiniae express multiple adhesion and invasion factors to colonize the intestinal tract. After the passage of the intestinal epithelial layer and entry into the underlying lymphatic tissues, they initiate a sophisticated host defense system to prevent their elimination by innate immune cells. Vice versa, the host triggers a complex multi-layered defense program. This includes a very strong inflammatory

Maria Kusmierek and Ann Kathrin Heroven contributed equally to this work.

Viveka Vadyvaloo and Matthew B. Lawrenz (eds.), *Pathogenic Yersinia: Methods and Protocols*, Methods in Molecular Biology, vol. 2010, https://doi.org/10.1007/978-1-4939-9541-7_8, © Springer Science+Business Media, LLC, part of Springer Nature 2019

and acute phase response, activation of the coagulation cascade for fibrin matrices formation around the bacterial colonies and initiation of a $T_H1/T_H17$ response to eliminate the bacteria [1]. During infection, however, multiple outcomes are possible: host cell death, bacterial persistence without symptoms, or full bacterial clearance [2, 3]. This suggests a heterogeneous programming of host–pathogen interactions during the course of the infection. To obtain a comprehensive understanding of the underlying processes, detailed knowledge of the associated gene expression profiles and their infection-linked changes during the distinct stages of the infection of both the host and the pathogen are required. Here, we describe a recently established probe-independent RNA-sequencing approach termed *Tissue Dual RNA-seq* that allows simultaneous profiling of a pathogen and its infected host to tackle this task [1, 3]. This very sensitive approach allowed us:

1. To reveal host immune responses that are triggered during the acute phase leading either to death of mice, or to clearance of the bacteria, or to the initiation of the persistence stage;
2. To discover virulence-relevant bacterial functions that are crucial for the survival of the pathogen in different host niches or during persistency that were not previously reported in the context of a *Yersinia* infection;
3. To identify regulatory factors as well as novel regulatory mechanisms and processes that drive different pathogenicity programs (early/late acute or persistent phase).

To this end, *Yersinia*-infected lymphatic tissue of the ileum (Peyer's patches) was lysed, total RNA of the bacterial and host cells was isolated and the ribosomal RNA of this RNA mixture was depleted. Subsequently, remaining RNA pools were converted into cDNA libraries and subjected to Illumina sequencing (Fig. 1). Sequencing reads were consecutively mapped to the *Yersinia* genome and the virulence plasmid, and the host core genome. Data normalization and analysis of the aligned reads allowed us to obtain highly accurate and sensitive transcriptional profiles of both the host and the pathogen from the data subsets. Moreover, a detailed pathway analysis combined with a differential expression analysis between in vitro grown bacterial cultures or uninfected mice has permitted identification of differentially expressed (a) bacterial and host genes during different stages of infection (acute versus chronic) or between wild-type or mutant infected mice (*Yersinia* wildtype versus *cnfY* mutant), and (b) numerous non-coding RNAs of the pathogen [1, 3]. The following detailed Tissue Dual RNA-seq protocol was established for the detection of bacterial (*Y. pseudotuberculosis* IP32953) and host (BALB/c mice) RNAs expressed in the Peyer's patches 3 days post-infection (Fig. 1).

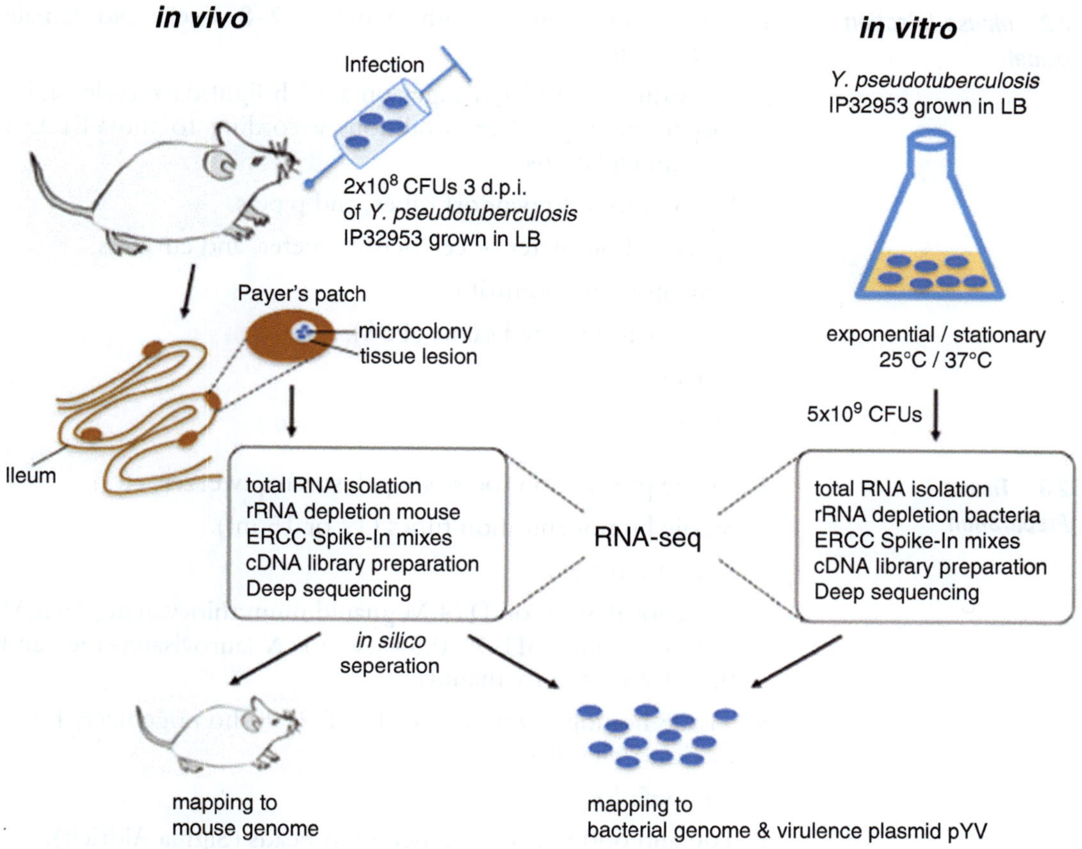


**Fig. 1** Tissue dual RNA-seq workflow. Five female BALB/c mice are orally infected with $2 \times 10^8$ CFUs of *Y. pseudotuberculosis* IP32953 (infected) or $1 \times$ PBS (uninfected) for *Tissue Dual RNA-seq.* For in vitro RNA-seq IP32953 is grown in LB to exponential or stationary phase at 25 °C or 37 °C. Total RNA is isolated from approximately $5 \times 10^9$ CFUs of in vitro grown *Y. pseudotuberculosis* IP32953. In parallel, total RNA is isolated from all isolated Peyer's patches of the infected and uninfected mice. The RNA is further processed for preparation of strand-specific barcoded cDNA libraries and sequenced. cDNA reads were separated in silico by mapping to the mouse genome mm10 and the *Y. pseudotuberculosis* IP32953 genome and the *Yersinia* virulence plasmid

## 2 Materials

### 2.1 Bacterial Cultures

1. *Yersinia pseudotuberculosis* IP32953 [4].
2. 100 ml Erlenmeyer flasks.
3. Shaker for Erlenmeyer flasks and incubator for bacterial plates.
4. Luria Bertani Broth (LB): 1% (w/v) tryptone, 0.5% (w/v) yeast extract, 85.6 mM sodium chloride; LB plates: LB broth, 1.8% (w/v) agar.

### 2.2 Mouse Infection Model

1. Six groups of five weight-matched 7–8 weeks old female BALB/c mice.
2. Individually ventilated cages on a 12 h light:dark cycle under specific pathogen-free conditions according to the FELASA recommendations.
3. Falcon tubes, Eppendorf tubes, and pipets.
4. Spectrophotometer or cell density meter, and cuvettes.
5. Benchtop microcentrifuge.
6. Phosphate-buffered saline (PBS).
7. Vortex.
8. Gavage needle.

### 2.3 Tissue Preparation

1. Tissue preparation tools (scalpel, scissor, tweezer, etc.).
2. Sterile homogenization tubes (12 or 15 ml).
3. Liquid nitrogen.
4. Pre-cooled Solution D (4 M guanidinium thiocyanate, 25 mM sodium citrate pH 7, 0.5% (w/v) *N*-lauroylsarcosine, and 0.1 M 2-mercaptoethanol).
5. Tissue homogenizer (Polytron PT 2100 homogenizer, Kinematica, Switzerland).
6. 2 ml safe-lock tubes.
7. 100 and 600 μm acid-washed glass beads (Sigma Aldrich).
8. TissueLyser (Qiagen).
9. 70% ethanol (EtOH).
10. Phosphate-buffered saline (PBS).
11. Benchtop microcentrifuge.

### 2.4 RNA Isolation, Stabilization and rRNA Depletion

1. 1.5 and 2 ml safe-lock tubes.
2. Resuspension buffer: 0.01 M sodium acetate (NaOAc) pH 4.5 and 0.3 M saccharose.
3. Lysis buffer: 0.01 M NaOAc pH 4.5 and 2% SDS.
4. Water-saturated phenol, pH 4.5–5.0.
5. Water bath or heat block.
6. Liquid nitrogen.
7. Benchtop microcentrifuge (refrigerated).
8. Vortex.
9. 3 M NaOAc pH 4.5.
10. 100% pure EtOH.
11. 70% EtOH, ice-cold.
12. Nuclease-free water.

13. TURBO DNase I (Thermo Fisher).
14. Phenol:chloroform:isoamyl alcohol (25:24:1) and chloroform: isoamyl alcohol (24:1).
15. Agilent RNA 6000 Nano Kit.
16. Agilent 2100 Bioanalyzer.
17. Ribo-Zero Gold rRNA Removal Kit Human/Rat/Mouse (Illumina).
18. MICROBExpress Bacterial mRNA Enrichment Kit (Thermo Fisher).
19. TE buffer, pH 8.0.
20. ERCC ExFold RNA Spike-In Mix 1 or Mix 2 (Thermo Fisher).

### 2.5 Library Preparation and Illumina Sequencing

1. Sonicator (Covaris Adaptive Focused Acoustics device, Covaris).
2. RNeasy MinElute Cleanup Kit (Qiagen).
3. Agilent RNA 6000 Nano Kit or Pico Kit, High Sensitivity DNA Kit.
4. Agilent 2100 Bioanalyzer.
5. T4 polynucleotide kinase (Thermo Fisher).
6. 3′- and 5′-RNA-adaptor oligonucleotides (for adapter sequences *see* Table 1, [5]).
7. T4 RNA ligase (Thermo Fisher).
8. SuperScript III Reverse Transcriptase (Invitrogen).
9. DNA primer complementary to the 3′-adapter sequence to generate first strand cDNA (for primers *see* Table 1, [5]).
10. Phusion® high fidelity DNA polymerase (NEB).
11. Polymerase chain reaction (PCR) primers which are identical to corresponding Illumina primers to ensure full compatibility with GenomeAnalyzer flowcells (Table 1, [5]).
12. QIAquick Gel Extraction Kit (Qiagen) or BluePippin (Biozym).
13. Capillary electrophoresis (Shimadzu MultiNA microchip electrophoresis system).
14. HiSeq2500 (Illumina).

### 2.6 Bioinformatics Tools

1. Genome Analyzer Pipeline Analysis software 1.8.2 (Illumina).
2. Fastq-mcf and fastq-multxtool of ea-utils [6] (http://hannonlab.cshl.edu/fastx_toolkit/).
3. *fastx_trimmer* from the FASTX-toolkit version 0.0.13.
4. *FastQC* program (http://www.bioinformatics.babraham.ac.uk/projects/fastqc/).

**Table 1**
**Oligonucleotides used for RNA sequencing**

| Name | Sequence (5′ > 3′) |
|---|---|
| *RNA adapters* | |
| Adapter-A-BC6-1 | UACACGACGCUCUUCCGAUCUACGAGA |
| Adapter-A-BC6-2 | UACACGACGCUCUUCCGAUCUACUUGC |
| Adapter-A-BC6-3 | UACACGACGCUCUUCCGAUCUAUCUGG |
| Adapter-A-BC6-4 | UACACGACGCUCUUCCGAUCUCAUACG |
| Adapter-A-BC6-5 | UACACGACGCUCUUCCGAUCUCGUCAU |
| Adapter-A-BC6-6 | UACACGACGCUCUUCCGAUCUGGUUAC |
| Adapter-A-BC6-7 | UACACGACGCUCUUCCGAUCUGUACCU |
| Adapter-A-BC6-8 | UACACGACGCUCUUCCGAUCUGUCAAG |
| Adapter-A-BC6-9 | UACACGACGCUCUUCCGAUCUUCUCUG |
| Adapter-A-BC6-10 | UACACGACGCUCUUCCGAUCUUGGUUC |
| Adapter-A-BC6-11 | UACACGACGCUCUUCCGAUCUUUCCGA |
| Adapter-A-BC6-12 | UACACGACGCUCUUCCGAUCUUUGCAG |
| Adapter-B[a] | P-AGAUCGGAAGAGCGGUUCAGC-ddC |
| *DNA primer* | |
| RT-primer | GCTGAACCGCTCTTCCGATCT |
| PCR-primer-A[b] | AATGATACGGCGACCACCGAGATCTACACTCTTTCCCTACACGACGC TCTTCCGATCT |
| PCR-primer-B[b] | CAAGCAGAAGACGGCATACGAGATCGGTCTCGGCATTCCTGC TGAACCGCTCTTCCGATCT |

[a] modifications of adapter-B were 5′-phosphate (P) and 3′-dideoxycytidine (ddC)
[b] adapted from Illumina PE PCR Primers 1 & 2. Original sequences were taken from http://intron.ccam.uchc.edu/groups/tgcore/wiki/013c0/Solexa_Library_Primer_Sequences.html

5. *Bowtie2* (version 2.1.0) [7].
6. *TopHat2* [8].
7. *SAMtools* [9].
8. *htseq-count* program [10].
9. *DESeq2* [11] (*see* **Note 1**).
10. KEGG database [12].
11. R package GOstats [13].
12. Signaling Pathway Impact Analysis (SPIA) algorithm [14].
13. ERCC RNA-Spike-In Control Mixes User Guide. https://tools.thermofisher.com/content/sfs/manuals/cms_086340.pdf.

14. TSSAR [15].
15. WebLogo software [16].
16. MEME software [17].
17. Integrated Genome Browser [18].

## 3 Methods

### *3.1 Mouse Infection with Y. pseudo-tuberculosis and Non-Infected Control Mice*

1. Prior to infection, starve 2 × 3 groups of five weight-matched BALB/c female mice aged between 7 and 8 weeks for several hours.
2. In parallel, grow *Y. pseudotuberculosis* strain IP32953 overnight ($OD_{600}$ approx. 4.0) in 20 ml LB broth in a 100 ml Erlenmeyer flask at 25 °C under aeration (2.24 × *g*) in a shaker.
3. Transfer a volume of 10 ml of the culture to a 50 ml reaction tube, fill up with sterile PBS to 50 ml and pellet the bacteria (2773 × *g*) for 10 min at 25 °C.
4. Remove the supernatant completely and wash the pellet again in 50 ml PBS.
5. Resuspend the bacteria in 10 ml PBS, determine the optical density and adjust the cells with PBS to $1 \times 10^9$ CFU/ml ($OD_{600} = 1$ corresponds to ca. $1 \times 10^9$ IP32953 CFU/ml).
6. Infect three groups of five mice intragastrically using a gavage needle with approximately $2 \times 10^8$ CFUs of *Y. pseudotuberculosis* IP32953 (200 μl/mouse) (*see* **Note 2**). For uninfected controls, treat three groups in parallel with PBS only.
7. Plate serial dilutions of the inoculum in triplicate to assess the actual CFU used for the infection.

### *3.2 Harvest of Infected and Non-Infected Lymphatic Tissue*

1. Sacrifice mice 3 days post infection, dissect, isolate, and transfer Peyer's patches of the infected and non-infected groups in precooled 2 ml test tubes and immediately snap freeze the samples in liquid nitrogen. Store at −80 °C or directly proceed with homogenization.

### *3.3 Homogenization of Peyer's Patches Tissue*

1. Cool homogenization tubes on ice.
2. Transfer the organ/tissue (Peyer's patches) from −80 °C freezer on ice.
3. Add 1.5 ml solution D to the organ and transfer to the homogenization tube.
4. Homogenize the tissue on ice at 1220 × *g* for 10 s. Wash the homogenizer before and in between tissues with 4 ml cold 70% EtOH and 4 ml PBS.
5. Transfer the tissue lysate to a 2 ml reaction tube.

6. In case of foamy samples, centrifuge the tissue lysate at 17,949 × *g* at 4 °C for 1 min.
7. Aliquot the lysate/supernatant in 2 × 0.75 ml to pre-cooled 2 ml safe-lock tubes containing 200 mg small (100 μm) and 200 mg big (600 μm) acid washed glass beads (i.e. 2 tubes for each sample).
8. Treat the sample with TissueLyser at fixed speed for 1 min (30 Hz) (*see* **Note 3**).
9. Cool on ice for 1 min.
10. Repeat **steps 8** and **9** two times.

### 3.4 Total RNA Isolation of Bacteria and Host Cells from Infected Tissue

1. Transfer 0.5 ml of the tissue lysate (without glass beads) to a new 2 ml reaction tube.
2. First, add 50 μl of 3 M sodium acetate pH 4.5 and mix the solution by inversion.
3. Add 500 μl water-saturated phenol and mix thoroughly by inversion.
4. Add 100 μl chloroform:isoamyl alcohol (24:1) and vortex the samples for 10 s, cool the samples on ice for 15 min.
5. Centrifuge the samples at 17,949 × *g* for 20 min at 4 °C.
6. Transfer the top layer to a new reaction tube and add 300 μl of chloroform:isoamyl alcohol (24:1). Vortex the mixture for 30 s and centrifuge the samples for 3 min at 17,949 × *g* at 4 °C.
7. Repeat **steps 4** and **5**.
8. Transfer the supernatant to a new 1.5 ml reaction tube and add 0.1 volumes of 3 M sodium acetate pH 4.5 and 2.5 volumes of 100% ice-cold EtOH. Invert the tubes several times and incubate the samples for at least 1 h at −20 °C to precipitate the RNA.
9. Pellet the RNA by centrifugation for 30 min at 17,949 × *g* at 4 °C.
10. Discard the supernatants and wash the pellets with 500 μl of 70% ice-cold EtOH.
11. Pellet the samples by centrifugation for 5 min at 17,949 × *g* at 4 °C.
12. Remove the supernatants and air-dry the pellets for 5–20 min at room temperature. Resuspend the pellets in 50 μl nuclease-free water.
13. Assess the final RNA concentration and RNA integrity with the Agilent RNA 6000 Nano Kit on the Agilent 2100 Bioanalyzer as described by the manufacturer.
14. Digest the residual DNA using TURBO DNase I (*see* Subheading 3.3, **step 4**) and analyze the RNA integrity again on the Agilent 2100 Bioanalyzer (Fig. 2) (*see* **Note 4**).

15. rRNA depletion and spiking of the RNA of the tissue samples.
    (a) Combine high-quality RNA (RIN ≥ 9) from Peyer's Patches of 5 mice to one RNA pool in equal amounts.
    (b) To reduce sequencing depth requirements, deplete mouse rRNA from three independent RNA pools using the Ribo-Zero Gold rRNA Removal Kit and purify the RNA by EtOH precipitation according to the instructions of the manufacturer.
    (c) For subsequent fragmentation, resuspend 800 ng RNA in TE buffer to reach the final volume of 130 μl (pH 8.0). Analyze the RNA integrity again on the Agilent 2100 Bioanalyzer (Fig. 2).
16. Add 1 μl of either 1:10 diluted ERCC ExFold RNA Spike-In Mix 1 or Mix 2 to 1 μg of mouse rRNA depleted RNA.

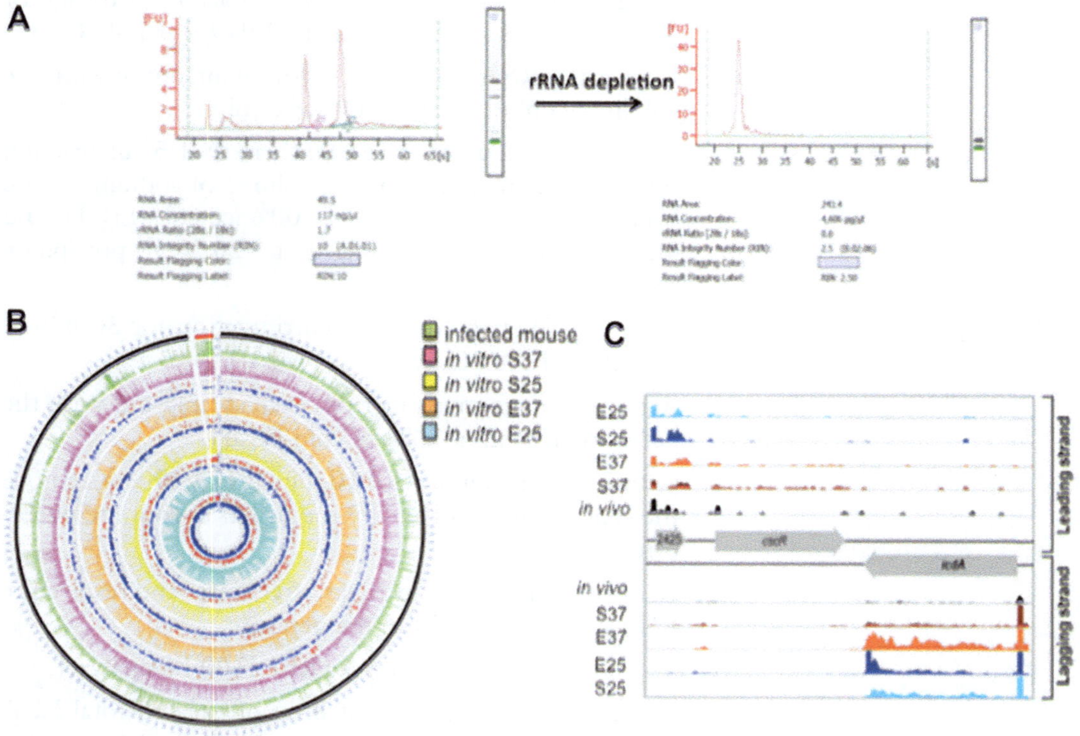


**Fig. 2** Global reports of the *Tissue Dual RNA-seq* analysis. (**a**) Representative Bioanalyzer profile of total RNA isolated from infected murine Peyer's patches before (left panel) and after (right panel) rRNA depletion. (**b**) Circos plot visualizing RPKM (Reads per Kilobase transcript length per million mapped reads) normalized expression values of in vitro and in vivo RNA-seq data for the IP32953 genome (NC_006155.1) and pYV virulence plasmid (NC_006153.2). (**c**) Read coverage of the RNA-seq analysis of the *cscR-icdA* locus is illustrated in the Integrated Genome Browser [18]. The data are normalized according to the number of uniquely mapped reads. E25: exponential phase 25 °C; E37: exponential phase 37 °C; S25: stationary phase 25 °C; S37: stationary phase 37 °C; in vivo: infected Peyer's patches

### 3.5 Total RNA Isolation from Bacterial Cultures In Vitro

1. Grow *Y. pseudotuberculosis* IP32953 in 20 ml LB broth in a 100 ml Erlenmeyer flask under aeration (2.24 × *g*) in a shaker in three separate cultures to exponential phase ($OD_{600}$ 0.5) or stationary phase ($OD_{600}$ approx. 4.0) at 25 °C or 37 °C.
2. Pellet the culture by centrifugation (2773 × *g* for 3 min at 25 °C or 37 °C) and use 5 × $10^9$ bacteria for total RNA preparation per growth condition and sample. Extract total RNA from bacterial pellets by the hot phenol extraction protocol described by Sambrook [19] (*see* **Note 5**).
3. To remove residual genomic and *Yersinia* virulence plasmid DNA remnants, treat the samples with 0.008 U of TURBO DNase I per 100 ng of RNA for 1 h at 37 °C in a total volume of 240 μl.
4. Perform a phenol–chloroform purification and EtOH precipitation with the DNA-free total RNA of the samples.
   (a) Add one volume of phenol:chloroform:isoamyl alcohol (25:24:1) to each sample. Mix the tubes by vortexing for 30 s and centrifuge for 3 min at 17,949 × *g* at 4 °C.
   (b) Extract the RNA containing supernatants twice with one volume of chloroform:isoamyl alcohol.
   (c) Transfer the aqueous phases into fresh 1.5 ml reaction tubes, supplemented with 0.1 volume of sodium acetate (pH 4.5) and 2.5 volumes of 100% ice-cold EtOH, and incubate the samples overnight at −20 °C to precipitate the RNA.
   (d) Pellet precipitated RNA by centrifugation for 30 min at 17,949 × *g* at 4 °C.
   (e) Remove the supernatants from each sample and wash the pellets with 1 ml of 70% ice-cold EtOH.
   (f) Centrifuge the samples for 10 min at 17,949 × *g*, remove the supernatants and air-dry the pellets at room temperature (RT).
   (g) Resuspend the pellets in RNase-free water and resolve the RNA for 15 min at room temperature.
   (h) Assess the RNA quality on Agilent 2100 Bioanalyzer.
5. To deplete ribosomal RNA, treat 8 μg of isolated total RNA with the MICROBExpress Bacterial mRNA Enrichment Kit according to the manufacturer's instructions.
6. For subsequent fragmentation, resuspend 800 ng RNA in TE buffer to reach the final volume of 130 μl (pH 8.0).
7. Add 1 μl of either 1:10 diluted ERCC ExFold RNA Spike-In Mix 1 or Mix 2 to 1 μg of rRNA depleted RNA.

### 3.6 cDNA Library Preparation

Modified from Dötsch et al. [5] (*see* **Note 6**).

1. For strand-specific RNA-seq cDNA library preparation and barcode introduction, perform a fragmentation of the rRNA-depleted RNA by sonication to a median size of 200 nt.
   (a) Fragment the RNA (from Subheadings 3.4, **step 11** and 3.5, **step 6**) to a median size of 200 nt by sonification with the following settings: processing time-150 s; fragment size range-200 bases; intensity-5; duty cyle-10%.
   (b) Precipitate the RNA by ethanol precipitation (*see* manual Ribo-Zero Gold rRNA Removal Kit) and resuspend the pellet in 16 μl nuclease-free water.
   (c) Use 1 μl for the subsequent analysis of quality and size distribution of the fragmented RNA on a Bioanalyzer Pico Chip.
2. To phosphorylate the 5′-ends for adapter ligation and to remove the 3′-phosphates, treat the remaining fragmented RNA with 1 μl (10 U) T4 polynucleotide kinase for 25 min at 37 °C. Perform phenol:chloroform purification and EtOH precipitation as described in Subheading 3.5, **step 4**. Resuspend the pellet in 16 μl nuclease-free water.
3. For the 3′ adapter ligation add 5 × concentration of 3′ adapter (*see* Table 1) from a 100 pmol/μl adaptor stock solution to RNA (e.g. 10 pmol RNA + 50 pmol adapter) and 1 μl of T4 RNA ligase.
4. Purify the RNA using the Qiagen RNeasy MinElute Cleanup Kit. Elute the RNA with 16 μl nuclease-free water.
5. For the 5′ adapter ligation add 5 × concentration of the 5′ adapters (*see* Table 1) from a 100 pmol/μl adaptor stock solution to RNA (e.g. 10 pmol RNA + 50 pmol adapter) and 1 μl of T4 RNA ligase.
6. Purify the RNA using the Qiagen RNeasy MinElute Cleanup Kit. Elute the RNA with 16 μl nuclease-free water.
7. For first strand cDNA synthesis add 1 μl RT primer (100 μM) (Table 1) to 4 μl RNA and perform PCR in thermocycler with the following program: 10 min 70 °C, followed by 10 min 25 °C. Place samples immediately on ice.
8. Prepare a mastermix for all samples: 2 μl 5 × first strand buffer; 1 μl 100 mM DTT; 1 μl 10 mM dNTP (each); 1 μl SuperScript III (per sample amounts).
9. Add the 5 μl of the mix to the 5 μl RNA/RT primer mix. Perform cDNA synthesis at 50 °C for 120 min.
10. To inactivate the reverse transcriptase, incubate the mix for 15 min at 70 °C.

11. To amplify the cDNA libraries perform a PCR reaction using Phusion polymerase: 10 μl of cDNA, 1 × Phusion polymerase HF buffer, 200 μM dNTPs, 0.5 μM primer A and primer B and 3% DMSO in a total volume of 50 μl. For thermocycling use the following program: 1. 98 °C 5 min; 2. 98 °C 1 min; 3. 62 °C 30 s; 72 °C 30 s; repeat **steps 2–4**: 15 cyles; 72 °C 5 min final elongation.
12. Load PCR products on a 2% thin agarose gel and cut out fragments (rather smear) between 150 and 500 bp.
13. Purify DNA using the QIAquick Gel Extraction Kit protocol or by automated DNA Size Selection with the BluePippin. Elute DNA with 30 μl nuclease-free water.
14. Use 1 μl of the samples for a subsequent analysis of quality and size distribution of the PCR products on the Bioanalyzer (High Sensitivity DNA Kit).

### 3.7 Illumina Sequencing

1. For Illumina HiSeq sequencing, combine all cDNA libraries in approximately equimolar amounts.
   (a) Apply a differential cleanup with Agencourt AMPure kit (Beckman Coulter Genomics) to fractionize cDNA pools in the range of 150–600 bp.
   (b) Analyze cDNA pools by capillary electrophoresis (Shimadzu MultiNA microchip electrophoresis system) (*see* **Note 7**).
2. The Illumina cluster station is used for cluster generation. Single-end sequencing on the HiSeq2500 was performed with 2 nM library denatured with 0.1 N NaOH and diluted with 5× SSC, 0.05% Tween-20 or HT-1 hybridization buffer of NextSeq 500 Kit (Illumina) to a final concentration of 12 pM. Cluster generation on HiSeqSR Flow Cell v3 was carried out at cBot using TruSeq SR Cluster Kit v3—HS to create single molecule DNA templates followed by bridge amplification. Sequencing run was performed at HiSeq 2500 using TruSeq SBS Kit v3 (50 cycle) to run 51 cycles and 7 cycles for the single-indexed read.

   Bar-coded 5′-adapters enable the pooling of multiple samples on one lane of the Illumina flow cell. Libraries are sequenced with 36–100 cycles in single-end mode, which resulted in 36–100 nt long reads including 6 nt of barcode and 30–94 nt of the sequenced 5′-ends of fragments (typically to > one million reads/library). Process the fluorescent images to sequences and transform the data to FastQ format using the Genome Analyzer Pipeline Analysis software 1.8.2 (Illumina).

### 3.8 Read Mapping and Initial Library Quality Assessment

1. Assess all FASTQ files from the sequenced libraries for sufficient read quality and potential contamination using the *FastQC* program (http://www.bioinformatics.babraham.ac.uk/projects/fastqc/).
2. Remove identified adapter contaminations and remaining artificial sequences (barcodes) from the Illumina reads in FASTQ format using program *fastx_trimmer* from the FASTX-toolkit version 0.0.13 (http://hannonlab.cshl.edu/fastx_toolkit/). At the 3′ end, trim reads if the per base Phred score falls short of 20. Discard trimmed reads with a remaining length < 20 nts.

### 3.9 Bioinformatic Analyses and Statistics

1. Map reads of the libraries to the *Y. pseudotuberculosis* IP32953 genome (NC_006155) and the pYV plasmid (NC_006153.2) using *Bowtie2* (version 2.1.0) [7] in end-to-end alignment mode with default parameterization. Employ *SAMtools* [9] to filter bam files for uniquely mapped reads (both strands). Reads are called uniquely mapped reads with a unique genomic location if and only if they could not be aligned to another location with a higher or same mapping quality. The uniquely mapped reads are classified as reads from *Yersinia*, whereas remaining non-mapping reads are categorized as mouse reads. Libraries with reads classified from mouse are aligned to the mouse genome (assembly: GRCm38/mm10) using the splice junction mapper *TopHat2* [8] with library type *fr-secondstrand* (*see* **Note 8**).
2. Potentially existing cross-mapping reads (very rare: < 0.001%, reads mapping to mouse and *Yersinia* for 50 bp single-end reads) identified by aligning the classified *Yersinia* reads back to the mouse genome using *Tophat2* [8] are removed from the libraries and excluded from downstream analyses (*see* **Note 9**).

### 3.10 Feature Normalization, Quantification and Detection of Differential Gene Expression

1. Quantify reads that align to annotated genes with program *htseq-count* [10]. For gene expression quantification in mouse use up-to-date gene annotations from the Ensembl database. For *Y. pseudotuberculosis* IP32953 use the custom annotation given in Dataset S6 in Nuss et al. [1]. This annotation enhances the NCBI annotation for NC_006155 and NC_006153.2 with 197 non-coding RNAs.
2. Use calibrated ERCC RNA spike-in-controls to estimate expression differences, assess the platform dynamic range and the accuracy of fold-change response [20] (*see* **Note 10**). Add spike-in control sequences to mouse reference genome/annotation prior to read alignment. Read counts for spike-in controls are determined along with normal gene counts with program *htseq-count*. For diagnostic plots of the dose- and fold-change-response follow the description of the manufacturer (ERCC RNA-Spike-In Control Mixes User Guide: https://tools.thermofisher.com/content/sfs/manuals/cms_086340.pdf).

3. Use Circos plot (Fig. 2b) to visualize RPKM (Reads per Kilobase transcript length per million mapped reads) of normalized expression values of in vitro and in vivo RNA-seq data for the IP32953 genome (NC_006155.1) and pYV virulence plasmid (NC_006153.2). Normalize data according to the number of uniquely mapped reads and illustrate read coverage of the RNA-seq analysis of selected loci with the Integrated Genome Browser [18] (Fig. 2c).
4. Use with *htseq-count* determined gene read counts as input for *DESeq2* [11] for pairwise detection and quantification of differential gene expression (*see* **Note 1**). Moreover, compute RPKM (reads per kilobase max. Transcript length per million mapped reads) values for each library from the raw gene counts, and filter the list of *DESeq2* generated differentially expressed genes (DEGs) (*see* **Note 11**). Mouse and *Yersinia* genes are classified as being differentially expressed if and only if $|\log2FC| \geq 2$ and (multiple testing corrected) $p$ value $\leq 0.05$ hold.

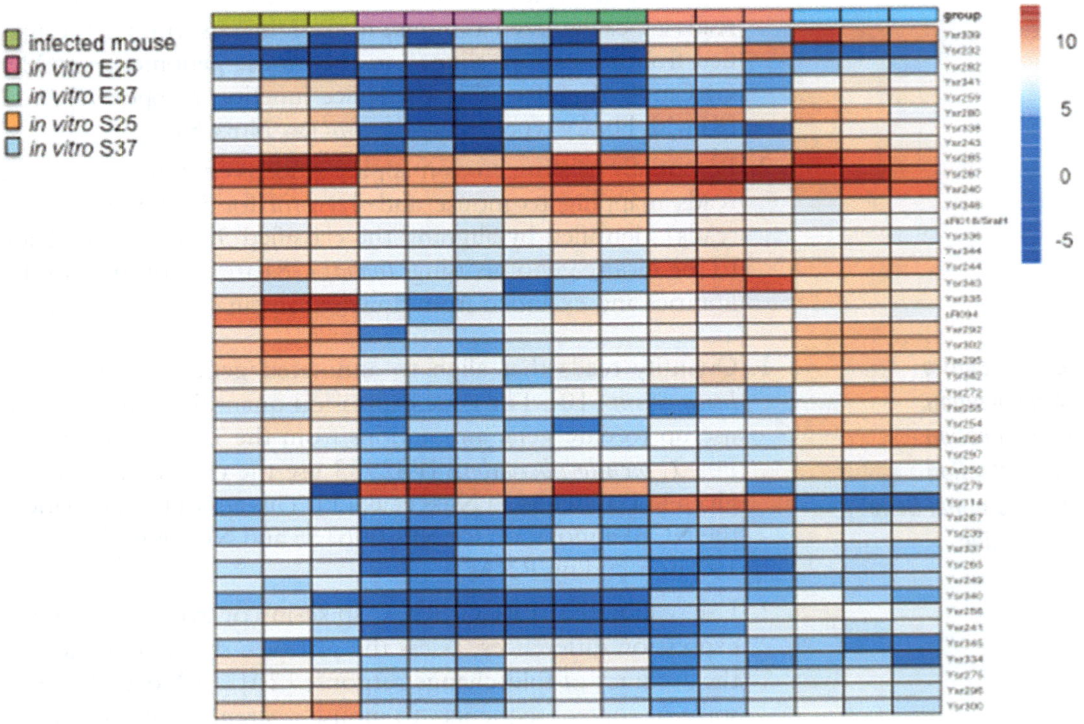


**Fig. 3** Comparison of regulatory/antisense RNAs of *Y. pseudotuberculosis* at different environmental conditions and during infection. Heatmap of the top enriched and depleted regulatory/antisense RNAs of the in vivo RNA-seq data from three independent RNA pools of the *Y. pseudotuberculosis* IP32953 core genome and the pYV virulence plasmid at different growth conditions. Color-coding is based on rlog-transformed read count values; enriched (red) or depleted (blue) during infection. E25: exponential phase 25 °C; E37: exponential phase 37 °C; S25: stationary phase 25 °C; S37: stationary phase 37 °C; in vivo: infected Peyer's patches

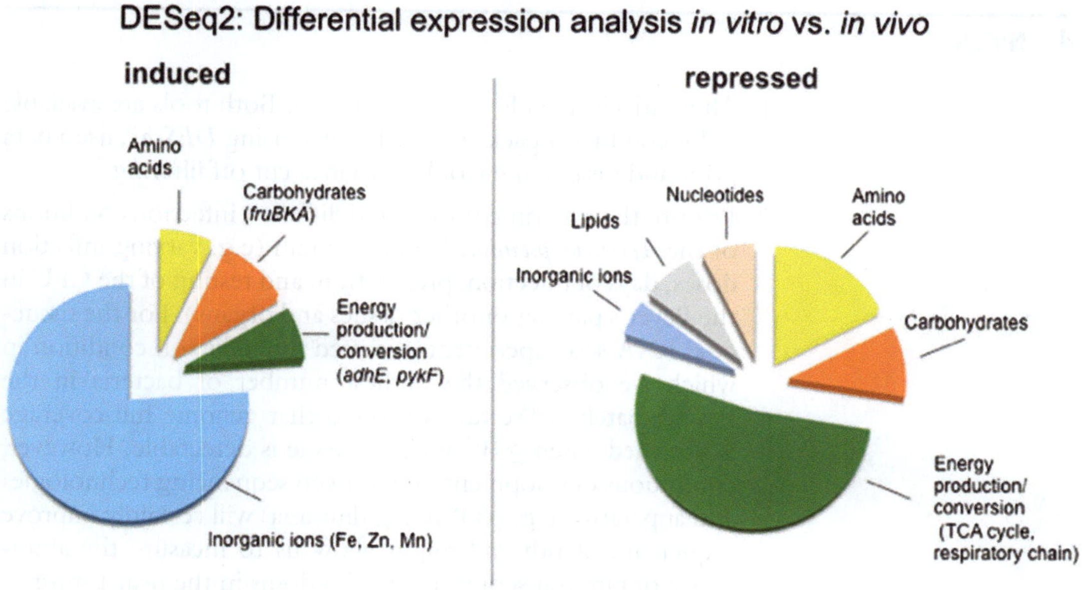


**Fig. 4** Bacterial global gene expression analysis uncovered infection-specific metabolic changes. Pie charts of metabolic functions, which are at least fourfold induced or repressed (*p* value 0.05) obtained by comparative RNA-seq using *DESeq2* from triplicate experiments (in vivo versus all four analyzed in vitro conditions)

### *3.11 Over-representation and Pathway Analyses*

1. Further annotate lists of *Yersinia* DEGs with pathway information from the KEGG database [12] and visualize by heat maps (e.g. differentially regulated antisense RNAs, Fig. 3) or pie charts (differentially regulated metabolic functions, Fig. 4).
2. Assess the list of DEGs resulting from comparison of infected mice and uninfected mice with functions from the R package GOstats [13] to obtain the association of Gene Ontology (GO) terms and KEGG metabolic pathways to genes. Use a *p* value cut-off of 0.001 for the applied conditional hypergeometric test for overrepresentation of GO terms in the three ontologies: molecular function, biological process, and cellular component and the annotated KEGG pathways.
3. Use GO annotations from the Bioconductor *Mus musculus* annotation package, and retrieve KEGG pathway annotations directly from KEGG using KEGG's REST API.
4. Apply the signaling pathway impact analysis algorithm [14] as implemented in the SPIA Bioconductor package with default parametrization to the list of mouse DEGs to identify host signaling processes influence by the *Yersinia* infection.

## 4 Notes

1. Alternatively, edgeR can be used [21]. Both tools are available as Bioconductor packages [22]. When using *DESeq2*, use a beta prior and disable the Cook's distance cut off filtering.
2. Prior to the experiment we tested different infection conditions of the *Yersinia pseudotuberculosis* strain (e.g. varying infection doses, days of infection, preparation, and testing of the CFU in the Peyer's patches or other tissues and organs). For the tissue-dual RNA-seq experiments we used the infection condition in which we observed the highest number of bacteria in the Peyer's patches. We further found that genome full-coverage is obtained when $\geq 10^6$ CFU/g tissue is detectable. However, continuous development of new deep sequencing technologies and apparatus (e.g. NOVAseq, Illumina) will certainly improve sequencing depth and might allow us to measure the abundance of rare transcripts across kingdoms in the near future.
3. We used the TissueLyser (Qiagen), but other mechanical lysis methods (e.g. using bead beating) or certain other lysis buffer systems (*mir*VanaKit, Thermo Fisher) could also be used for tissue lysis, but they need to be optimized for the individual *Yersinia* strain.
4. Proceed with the following steps only if the RNA quality is high (RIN $\geq$ 9).
5. It is extremely important to test different experimental set-ups for RNA extraction. We note that the hot phenol extraction protocol worked best for us without addition of RNA stabilizer. In particular, when tissue samples are immediately processed, addition of RNA stabilizer seemed expendable. Nonetheless, different RNA stabilization agents were tested, for example, RNA*later* and RNA*protect* (Qiagen). *Yersinia pseudotuberculosis* growth was immediately stopped in RNA*later*, but the bacteria survived the procedure and started to regrow when the sample was diluted or the buffer was exchanged. Moreover, the overall yield of the RNA was reduced and it is also important to know that it quenched fluorescent signals of Gfp-expressing bacteria in case a sorting process is planned to isolate bacteria from the tissue. Compared to RNA*later*, RNA*protect* was less efficient in *Yersinia* RNA stabilization.
6. We like to note that our protocol for cDNA library was established for sequencing on the HiSeq2500 (Illumina), but currently new technologies, that is, NOVAseq is used and protocols need to be optimized accordingly.
7. Proceed with Illumina HiSeq sequencing only if the fragments are distributed in the range of 150–600 bps.

8. Alternatively, other splice aware alignment programs like for example, STAR [23] can be used here.
9. Three different categories of mouse and *Yersinia* read sets are classified based on their origin: the infected (IP) or uninfected (PBS) mouse read set, and the *Yersinia* read set from the infected mouse samples. Moreover, four *Yersinia* read sets are generated from the in vitro grown pools as a reference (25 °C and 37 °C exponential phase, 25 °C and 37 °C stationary phase). The resulting bam files of these read sets constitute the basis for quantification, comparison, and visualization.
10. Spike-ins should be added as early as possible to avoid biases in down-stream processing steps.
11. We successively aligned obtained reads to the *Yersinia* and then to the mouse genome. However, also parallel alignment to respective reference genomes is possible, and other pipelines can be used [24]. Moreover, reads cross-mapping to both, the pathogen and the host genome were identified and discarded although the overall proportion was very small to optimize differential expression analysis.

## Acknowledgments

We are grateful to M. Fenner for discussions and Robert Geffers and Michael Jarek from the Department of Genome Analytics for Illumina sequencing. This work was supported from grants of the German Research Foundation (DE616/4, DE616/6, SPP1617-young investigator startup funding for A.M. Nuss), and a stipend of the Helmholtz Center for Infection Research Graduate School for M. Kusmierek. P. Dersch is supported by the German Center for Infection Research.

## References

1. Nuss AM, Beckstette M, Pimenova M, Schmühl C, Opitz W, Pisano F, Heroven A, Dersch P (2017) Tissue dual RNA-seq: a fast discovery path for infection-specific functions and riboregulators shaping host-pathogen transcriptomes. Proc Natl Acad Sci U S A 114 (5):E791–E800
2. Avican K, Fahlgren A, Huss M, Heroven AK, Beckstette M, Dersch P, Fallman M (2015) Reprogramming of *Yersinia* from virulent to persistent mode revealed by complex in vivo RNA-seq analysis. PLoS Pathog 11(1): e1004600. https://doi.org/10.1371/journal.ppat.1004600
3. Heine W, Beckstette M, Heroven AK, Thiemann S, Heise U, Nuss AM, Pisano F, Strowig T, Dersch P (2018) Loss of CNFY toxin-induced inflammation drives *Yersinia pseudotuberculosis* into persistency. PLoS Pathog 14(2):e1006858. https://doi.org/10.1371/journal.ppat.1006858
4. Chain PS, Carniel E, Larimer FW, Lamerdin J, Stoutland PO, Regala WM, Georgescu AM, Vergez LM, Land ML, Motin VL, Brubaker RR, Fowler J, Hinnebusch J, Marceau M, Medigue C, Simonet M, Chenal-Francisque V, Souza B, Dacheux D, Elliott JM, Derbise A, Hauser LJ, Garcia E (2004) Insights into the evolution of *Yersinia pestis* through whole-genome comparison with *Yersinia pseudotuberculosis*. Proc Natl Acad Sci U S A 101 (38):13826–13831
5. Dotsch A, Eckweiler D, Schniederjans M, Zimmermann A, Jensen V, Scharfe M,

Geffers R, Haussler S (2012) The *Pseudomonas aeruginosa* transcriptome in planktonic cultures and static biofilms using RNA sequencing. PLoS One 7(2):e31092. https://doi.org/10.1371/journal.pone.0031092

6. Aronesty E (2011) ea-utils: Command-line tools for processing biological data
7. Langmead B, Salzberg SL (2012) Fast gapped-read alignment with bowtie 2. Nat Methods 9 (4):357–359. https://doi.org/10.1038/nmeth.1923
8. Kim D, Pertea G, Trapnell C, Pimentel H, Kelley R, Salzberg SL (2013) TopHat2: accurate alignment of transcriptomes in the presence of insertions, deletions and gene fusions. Genome Biol 14(4):R36. https://doi.org/10.1186/gb-2013-14-4-r36
9. Li H, Handsaker B, Wysoker A, Fennell T, Ruan J, Homer N, Marth G, Abecasis G, Durbin R, Genome Project Data Processing S (2009) The sequence alignment/map format and SAMtools. Bioinformatics 25 (16):2078–2079. https://doi.org/10.1093/bioinformatics/btp352
10. Anders S, Pyl PT, Huber W (2015) HTSeq--a Python framework to work with high-throughput sequencing data. Bioinformatics 31(2):166–169. https://doi.org/10.1093/bioinformatics/btu638
11. Love MI, Huber W, Anders S (2014) Moderated estimation of fold change and dispersion for RNA-seq data with DESeq2. Genome Biol 15(12):550. https://doi.org/10.1186/s13059-014-0550-8
12. Kanehisa M, Goto S, Sato Y, Kawashima M, Furumichi M, Tanabe M (2014) Data, information, knowledge and principle: back to metabolism in KEGG. Nucleic Acids Res 42 (Database issue):D199–D205. https://doi.org/10.1093/nar/gkt1076
13. Falcon S, Gentleman R (2007) Using GOstats to test gene lists for GO term association. Bioinformatics 23(2):257–258. https://doi.org/10.1093/bioinformatics/btl567
14. Tarca AL, Draghici S, Khatri P, Hassan SS, Mittal P, Kim JS, Kim CJ, Kusanovic JP, Romero R (2009) A novel signaling pathway impact analysis. Bioinformatics 25(1):75–82. https://doi.org/10.1093/bioinformatics/btn577
15. Amman F, Wolfinger MT, Lorenz R, Hofacker IL, Stadler PF, Findeiss S (2014) TSSAR: TSS annotation regime for dRNA-seq data. BMC Bioinformatics 15:89. https://doi.org/10.1186/1471-2105-15-89
16. Crooks GE, Hon G, Chandonia JM, Brenner SE (2004) WebLogo: a sequence logo generator. Genome Res 14(6):1188–1190. https://doi.org/10.1101/gr.849004
17. Bailey TL, Elkan C (1994) Fitting a mixture model by expectation maximization to discover motifs in biopolymers. Proc Int Conf Intell Syst Mol Biol 2:28–36
18. Nicol JW, Helt GA, Blanchard SG Jr, Raja A, Loraine AE (2009) The integrated genome browser: free software for distribution and exploration of genome-scale datasets. Bioinformatics 25(20):2730–2731. https://doi.org/10.1093/bioinformatics/btp472
19. Sambrook J (2001) Molecular cloning: a laboratory manual. Cold Spring Harbor Laboratories, Cold Spring Harbor, NY
20. Jiang L, Schlesinger F, Davis CA, Zhang Y, Li R, Salit M, Gingeras TR, Oliver B (2011) Synthetic spike-in standards for RNA-seq experiments. Genome Res 21(9):1543–1551. https://doi.org/10.1101/gr.121095.111
21. Robinson MD, McCarthy DJ, Smyth GK (2010) edgeR: a bioconductor package for differential expression analysis of digital gene expression data. Bioinformatics 26(1):139–140. https://doi.org/10.1093/bioinformatics/btp616
22. Huber W, Carey VJ, Gentleman R, Anders S, Carlson M, Carvalho BS, Bravo HC, Davis S, Gatto L, Girke T, Gottardo R, Hahne F, Hansen KD, Irizarry RA, Lawrence M, Love MI, MacDonald J, Obenchain V, Oles AK, Pages H, Reyes A, Shannon P, Smyth GK, Tenenbaum D, Waldron L, Morgan M (2015) Orchestrating high-throughput genomic analysis with bioconductor. Nat Methods 12(2):115–121. https://doi.org/10.1038/nmeth.3252
23. Dobin A, Davis CA, Schlesinger F, Drenkow J, Zaleski C, Jha S, Batut P, Chaisson M, Gingeras TR (2013) STAR: ultrafast universal RNA-seq aligner. Bioinformatics 29(1):15–21. https://doi.org/10.1093/bioinformatics/bts635
24. Westermann AJ, Forstner KU, Amman F, Barquist L, Chao Y, Schulte LN, Muller L, Reinhardt R, Stadler PF, Vogel J (2016) Dual RNA-seq unveils noncoding RNA functions in host-pathogen interactions. Nature 529 (7587):496–501. https://doi.org/10.1038/nature16547

Chapter 9

# Detection of Cells Translocated with *Yersinia* Yops in Infected Tissues Using β-Lactamase Fusions

**Giang T. Nguyen, Anne L. McCabe, Alyssa C. Fasciano, and Joan Mecsas**

## Abstract

Development of the TEM-CCF2/4-AM FRET-based system has enabled investigators to track translocation of effector proteins into mammalian cells during infection. This allows for separation of translocated and non-translocated cell populations for further study. *Yersinia* strains expressing translational Yop-TEM fusions, containing the secretion and translocation signals of a Yop with the TEM-1 portion of β-lactamase, are used to infect mice, tissues isolated from mice, or mammalian cells in culture. Infected and harvested mammalian cells are treated with either CCF2-AM or CCF4-AM, and cleavage of this fluorescent compound by TEM is detected by fluorescence-activated cell sorting (FACS) analysis. A shift from green to blue emission spectra of individual cells is indicative of translocation of a given Yop-TEM fusion protein into the host cell during *Yersinia* infection due to a disruption in FRET between the two fluors of the compound. In *Yersinia*, this method has been used to understand Type III secretion dynamics and Yop functions in cells translocated by effectors during infection. Here, we describe how to generate Yop-TEM constructs, and how to detect, quantify, isolate, and study Yop-TEM containing cells in murine tissues during infection and in ex vivo tissues by cell sorting and flow cytometry analysis. In addition, we provide guidance for analyzing TEM-positive cells via a plate reader and fluorescent microscopy.

**Key words** β-Lactamase, Bla, TEM, Yop translational fusions, *Yersinia*, Yops, Translocation, Neutrophils, CCF2, CCF4, FACS, T3SS, Type III secretion, Flow cytometry

## 1 Introduction

All three *Yersinia* species use a virulence plasmid-encoded Type III secretion system to translocate bacterial effector proteins (Yops) into eukaryotic cells [1–4]. These Yops have wide-ranging effects on host cellular functions and allow bacteria to evade host immune responses to establish infection [4–6]. For many years it had been challenging to identify and study the host cells injected with Yops during tissue infection, in part because *Yersinia* species are extracellular. Furthermore, it can be difficult to track Yop translocation into individual host cells by traditional fluorescence or epitope tagging methods without having to fix or lyse host cells. The development of the TEM-CCF2/4 system [7, 8] has allowed

Viveka Vadyvaloo and Matthew B. Lawrenz (eds.), *Pathogenic Yersinia: Methods and Protocols*, Methods in Molecular Biology, vol. 2010, https://doi.org/10.1007/978-1-4939-9541-7_9, © Springer Science+Business Media, LLC, part of Springer Nature 2019

investigators to identify host cells targeted for injection by Yops by all three pathogenic *Yersinia* species and characterize both effector translocation and host responses to infection [9–11].

The Yop-TEM-CCF2/4 systems work by creating a translational fusion protein with the secretion and translocation signal sequences of Yops (contained within the first 100 amino acids of each Yop) [12] to the TEM-1 portion of β-lactamase. TEM-1 lacks the N-terminal signal sequence (23 amino acids) of β-lactamase [13] but retains its enzymatic ability to hydrolyze beta-lactam bonds in antibiotics and other molecules [13, 14]. Upon translocation into mammalian cells, TEM cleaves the β-lactam ring of the cephalosporin core of the substrates CCF2 and CCF4. CCF2-AM and CCF4-AM are membrane-permeable substrates that contain two fluorophores linked by a cephalosporin ring. Once added to mammalian cells, CCF2-AM and CCF4-AM are cleaved by endogenous cytoplasmic esterases into their negatively charged forms and are retained in the cytosol. In the absence of TEM, excitation of the donor fluor (7-hydroxycoumarin) at 409 nm leads to the FRET excitation of the acceptor fluor (fluorescein) at 520 nm, and emission of green light. When TEM cleaves the cephalosporin β-lactam ring and separates the two fluors, excitation of 7-hydroxycoumarin at 409 nm leads to emission of blue light at 447 nm and the quenching of the acceptor fluor [8]. Thus, shifts from green to blue emissions from mammalian cells containing CCF2/4-AM after infection with a strain expressing a Yop-TEM fusion are a direct measurement of translocation of that effector protein into the host cells. Methods used to detect shifts from blue to green fluorescence include fluorescence activated cell sorting (FACS), plate readers, and microscopy.

The Yop-TEM CCF2/4-AM FRET system is powerful because investigators can identify, separate, and study translocated cells from non-translocated cells in infected tissues. This method has been used in all three pathogenic *Yersinia* species to show the range of cells injected with Yops in different tissues during various types of infection including oral, intravenous, and intranasal infection [9–11, 15–17]. Investigators have further exploited this method to study cells isolated from infected tissues and examine the consequences of Yop injection [18]. Furthermore, bacterial and/or host mutants can be used to determine the contributions of specific Yop effectors, bacterial ligands, and host factors in modulating *Yersinia*-host cell interactions in infected tissues [15, 17, 19–24]. Finally, this method has been used in high-throughput screens to identify small molecule inhibitors of Yop translocation [25, 26].

In this chapter, we describe construction of Yop-TEM translational fusions, which are expressed either on plasmids [9] or integrated into *Yersinia* genomic DNA [10]. We also describe analytical FACS methods to identify and quantify cells translocated with Yops [9–11, 16, 17] and steps subsequent to FACS to analyze the effects of Yop translocation in mammalian cells from infected tissues [18]. In addition, while not the primary focus of this chapter, we describe methods for plate reader assays for Yop-TEM translocation measurements [25, 26].

## 2 Materials

### 2.1 Construction of Yop-TEM Plasmids

1. Cloning plasmids: pBR322 (contains *blaM* gene), donor plasmid of interest (*see* **Note 1**).
2. Cloning reagents: restriction enzymes, CutSmart Buffer.
3. DNA oligonucleotides (*see* **Note 2** for design).
4. PCR reagents: Phusion polymerase, 5× Phusion buffer, 50 mM $MgCl_2$, 100 μM dNTPs.
5. Ligation reagents: T4 DNA ligase, T4 DNA ligase buffer.
6. SY327 or DH5α λpir competent cells.
7. Luria Media Broth (LB) broth.
8. LB agar plates.
9. Kanamycin or antibiotic to select for plasmid.

### 2.2 Bacterial Conjugation

1. SY327 or DH5α λpir.
2. *Escherichia coli* strain with pRK600 donor plasmid.
3. *Yersinia* recipient strain of interest.
4. 2× YT broth.
5. LB agar plates.
6. LB-irgasan/kanamycin plates.
7. Toothpicks.
8. Kanamycin.
9. Chloramphenicol.
10. Antibodies against β-lactamase.

### 2.3 Verification of TEM Constructs

1. Trichloroacetic acid (TCA).
2. Acetone.
3. SDS sample buffer.
4. SDS-PAGE gels and supplies for running western blots.
5. Coomassie stain.
6. Antibodies to beta lactamase.

### 2.4 Preparation of Yersinia

1. Small culture tubes.
2. 2× YT media for overnight *Yersinia* cultures.

3. Low $Ca^{2+}$ 2× YT: 2× YT with 20 mM $Na_2C_2O_4$, 20 mM $MgCl_2$.
4. Optical density spectrophotometer.
5. Cuvettes.
6. 1× PBS.
7. DMEM.
8. RPMI 1640.
9. Heat-inactivated fetal calf serum (HI-FBS) (*see* **Note 3**).
10. LB-Irgasan plates.

### 2.5 Preparation of 10× CCF2/4-AM

1. 10× CCF2/4-AM for tissue culture experiments:1 μg/mL CCF2/4-AM, 1.5 mM probenecid, and 100 μg/mL gentamicin (final concentrations). To make a 1 mM stock solution of CCF2/4-AM, add 185 μL of dimethyl sulfoxide (DMSO) to 200 μg CCF2-AM or 182 μL of DMSO to 200 μg CCF4-AM. Store at −20 °C in 6 μL aliquots. To prepare 600 μL of a 10× CCF2/4-AM stock mixture, add 36 μL solution B (provided with CCF2/4) to 6 μL 1 mM CCF2/4 in DMSO. Vortex, add 36 μL probenecid acid (250 mM stock); vortex. Add 462 μL PBS, vortex, add 60 μL gentamicin (10 mg/mL stock).
2. 10× CCF2/4 for ex vivo and in vivo mouse infections: 1 μg/mL CCF2/4-AM, 1.5 mM probenecid (final concentrations). To make a 1 mM stock solution of CCF2/4-AM, add 185 μL of DMSO to 200 μg CCF2-AM or 182 μL of DMSO to 200 μg CCF4-AM. Store at −20 °C in 6 μL aliquots. To prepare 600 μL of a 10× CCF2/4-AM stock mixture, add 36 μL solution B (provided with CCF2/4) to 6 μL 1 mM CCF2/4 in DMSO. Vortex, add 36 μL probenecid acid (250 mM stock); vortex. Add 522 μL PBS, vortex.
3. Aluminum foil.

### 2.6 In Vitro Infection of Cultured Cells

1. 24-well plate.
2. Confluent cultured cells.
3. Tissue culture media (e.g. DMEM +10% HI FBS).
4. 0.25 M Trypsin.
5. 1× PBS.
6. 5 mL polystyrene tubes for LSRII/FACS analysis.
7. LB-Irgasan plates.
8. Optical density spectrophotometer.
9. Cuvettes.
10. 70 μm filter.

### 2.7 Ex Vivo Infection of Splenocytes or Lung Cells

1. Optical density spectrophotometer.
2. Cuvettes.
3. RPMI 1640.
4. RPMI 1640 + 5% HI-FBS.
5. 2 mL Eppendorf tubes.
6. 70-μm cell strainer.
7. 5 mL syringes.
8. 6-well plate.
9. 1 mg/mL of Collagenase D dissolved in PBS or water.
10. Pharm Lyse 10× stock; dilute in sterile $H_2O$ for 1×.
11. 24-well plate.
12. 30-gauge needle and syringe.

### 2.8 Mouse Infection

1. Optical density spectrophotometer.
2. Cuvettes.
3. 1× PBS.
4. Anesthetics.
5. Anesthesia chamber.
6. 50% glycerol.
7. 1× PBS, $Ca^{2+}$- and $Mg^{2+}$-free.
8. 5 mL syringes.
9. 6-well plates.
10. RPMI 1640.
11. 70-μm cell strainer.
12. FACS buffer: 1% FBS in 1× PBS, $Ca^{2+}$- and $Mg^{2+}$-free.
13. 1× Fc blocking solution: 10 μL Purified Rat α-mouse CD16/CD32 (0.5 mg/mL stock concentration) in 2 mL of FACS buffer.
14. Antibodies to surface markers.
15. 5 mL polypropylene tubes for MoFlo/FACS sorter.
16. 5 mL polystyrene tubes for LSRII/FACS analysis.
17. LSRII (Becton Dickson), or similar FACS machine equipped with a 409 nm laser, and 447 nm and 520 nm filters.

### 2.9 Preparation of TEM⁺ and TEM⁻ Neutrophils (PMNs)

1. FACS buffer: 1% FBS in 1× PBS, $Ca^{2+}$- and $Mg^{2+}$-free.
2. Sorting buffer: FACS buffer, 5 mM EDTA, and 50 mM HEPES.
3. 1× $Ca^{2+}$-, $Mg^{2+}$-, and phenol red-free Hank's Balanced Salt Solution (HBSS).

4. 1× PBS, $Ca^{2+}$-, $Mg^{2+}$-free.
5. 1× phenol red-free HBSS with $Ca^{2+}$ and $Mg^{2+}$.
6. 1 M glucose.

#### 2.10 Analysis by Microscopy

1. Coumarin/Pacific Blue Longpass Filters (Chroma, Set Number: 19011).
2. 10× or 40× Objective.
3. Either epifluorescence or confocal microscope.

#### 2.11 RT-PCR

1. FACS buffer: 1% FBS in 1× PBS, $Ca^{2+}$- and $Mg^{2+}$-free.
2. Sorting buffer: FACS buffer, 5 mM EDTA, and 50 mM HEPES.
3. TRIzol® Reagent.
4. RNA extraction kits (e.g. QIAGEN RNeasy Mini Kit).

#### 2.12 Cytokine Analysis

1. RPMI 1640.
2. Fetal bovine serum (FBS).
3. Brefeldin A (BFA) – refer to manufacturer's instructions, but stock concentration is typically at 10 mg/mL. Store at −20 °C.
4. BD cytometric bead array or specific fluorescently labeled antibodies for cytokines of choice.
5. Buffers for intracellular staining – 4% formaldehyde to fix cells; FACS buffer: 1% FBS in 1× PBS, $Ca^{2+}$- and $Mg^{2+}$-free; and permeabilization buffer: 0.1% saponin in FACS buffer for temporary permeabilization of cells.

#### 2.13 Analysis by Plate Reader

1. Tissue culture media (e.g. DMEM with 10% FBS or RPMI with 5% FBS).
2. 384-well plates.
3. Gentamicin 10 mg/L stock in $H_2O$.

## 3 Methods

#### 3.1 Generation of Yop-TEM Translational Fusion Construct Plasmid

1. Polymerase chain reaction (PCR) amplify the *blaM* gene lacking the signal sequence (TEM) using primers containing restriction sites compatible with the donor plasmid (*see* **Note 2**).
2. Ligate TEM into donor plasmid using standard molecular biology techniques.
3. Transform into DH5α λpir and select on antibiotic agar plates.
4. PCR amplify the promoter and the desired portion of the Yop of interest containing the secretion and translocation signals

(usually the first 100 amino acids) from *Yersinia* DNA with primers containing restriction sites compatible with the multiple cloning site (MCS) upstream from the TEM insertion into the donor plasmid (*see* **Notes 4** and **5**).

5. Ligate the Yop promoter upstream of TEM in the donor plasmid using standard molecular biology techniques.
6. Transform into DH5α λpir and select on antibiotic agar plates.

### 3.2 Generation of Yersinia Strains Expressing Yop-TEM Translational Fusions

Depending on the plasmid, the Yop-TEM translational fusion construct can be electroporated into *Yersinia* to be expressed *in trans*, or it can be integrated into the chromosome by allelic exchange as described here.

#### 3.2.1 Conjugation of Yop-TEM Translational Fusion Construct into Yersinia

1. For allelic exchange, subclone the Yop-TEM translational fusion construct into a plasmid containing regions of homology to the desired *Yersinia* chromosomal insertion site to generate a merodiploid using standard molecular biology techniques (*see* **Note 6**).
2. Inoculate 2 mL cultures of three strains: recipient strain (*Yersinia*) in 2× YT, helper strain (DH5α with pRK600 with 40 μg/mL chloramphenicol) and donor strain (SY327 or DH5α λpir with plasmid of interest) in LB broth with appropriate antibiotics. Grow cultures overnight (*see* **Note** 7).
3. Transfer 500 μL of each overnight culture in new Eppendorf tubes, and spin at 17,000 × *g* for 2 min. Remove supernatant, add 500 μL antibiotic-free 2× YT and spin at 17,000 × *g*. Repeat twice.
4. Remove supernatant and resuspend the pellet in 500 μL 2× YT without antibiotics.
5. In 4 spots on an LB plate, place 5 μL of each strain on top of each other. Also spot each strain (5 μL) individually on the bottom part of the plate, which will serve as negative controls for selection on drug plates.
6. Invert once liquid is absorbed and leave at room temperature (RT) overnight.
7. In the morning use a toothpick, scoop up entire mating spot and smear on antibiotic selective plate (LB-Irgasan + antibiotic appropriate for plasmid selection: Kanamycin in case of pRS47), one spot per plate. Include individual strain spot controls.
8. In the afternoon, scoop up entire smeared field from each plate and streak for isolated colonies on new selective plates. Let *Yersinia* grow at RT for 2 days. Control colonies (individual strains) should not grow on selective plates.
9. Pick two isolated colonies from each mating and re-streak for isolated colonies on selective plates. Incubate at RT for 2 days.

10. Take an isolated colony from each streak and grow in 2× YT overnight at 26 °C with antibiotics for trichloroacetic acid (TCA) precipitation (*see* below).

#### 3.2.2 Verification of Secretion of Yop-TEM Fusion Protein

Test for Yop-TEM secretion by trichloroacetic acid (TCA) precipitation assay of supernatants followed by Western Blot analysis with an antibody against Beta-lactamase.

1. Day 1 afternoon: Prepare overnight cultures of *Yersinia* for infection: Inoculate single colonies into individual sterile tubes containing 2 mL 2× YT plus antibiotics with *Yersinia* Yop-TEM translational fusion strain. Incubate overnight at 26 °C with aeration. For screening colonies from mating, we recommend inoculating 8–10 single colonies.
2. Day 2: Dilute overnight *Yersinia* cultures 1:40 in 2 mL low $Ca^{2+}$ 2× YT (*see* **Notes 8** and **9**).
3. Grow *Yersinia* for 2 h shaking at 26 °C, and then shift to 2 h shaking at 37 °C.
4. Transfer 900 μL of bacterial culture to new Eppendorf tube. Add 100 μL TCA to each tube for a final concentration of 10%.
5. Incubate on ice for 15 min.
6. Spin down at max speed (17,000 × *g*) for 15 min.
7. Carefully decant supernatant; avoid aspirating pellet.
8. Wash pellet with ice-cold acetone (500 μL), and spin down max speed for 5 min.
9. Carefully decant supernatant and air-dry pellet.
10. Resuspend pellet in 50 μL SDS sample buffer, boil samples at 95 °C for 5 min, and run 10 μL on SDS-PAGE gel and stain with Coomassie to visualize secreted Yops.
11. Confirm the Yop-TEM fusion protein is secreted by Western blot using TEM-specific antibodies.
12. Freeze down (−80 °C) a single colony of the *Yersinia* Yop-TEM translational fusion strain that secretes Yops properly in 2× YT with 20% glycerol.

### 3.3 In Vitro Infection of Cultured Cells with Yersinia Yop-TEM Translational Fusion Construct

#### 3.3.1 Infection of Cultured Cells with Yersinia

1. Day 1: Plate cells in a 24-well plate (*see* **Note 10**).
2. Day 1: afternoon – Prepare overnight cultures of *Yersinia* for infection: Inoculate 2 mL 2× YT plus antibiotics with *Yersinia* Yop-TEM translational fusion strain. Incubate overnight at 26 °C with aeration.
3. Day 2: Dilute overnight *Yersinia* cultures 1:40 in 2 mL low $Ca^{2+}$ 2× YT (*see* **Notes 8** and **9**).
4. Grow *Yersinia* for 2 h shaking at 26 °C, and then shift to 2 h shaking at 37 °C.

5. Take $OD_{600}$ of cells to estimate culture concentration, and dilute in DMEM (or appropriate cell culture media) to desired concentration.
6. Plate serial dilutions of *Yersinia* used in infection on LB-Irgasan plates to determine multiplicity of infection (MOI). Grow at 26 °C for 2 days before counting CFUs.
7. Add 100 μL of *Yersinia* in DMEM at the desired MOI (based on starting cell concentration) to each well (*see* **Note 11**).
8. Spin 3 min at 20 × *g* at RT.
9. Incubate 1–2 h in tissue culture incubator or until cells round up (*see* **Note 12**).

#### 3.3.2 Preparation of Samples for FACS Analysis

1. Remove media from each well. Wash once with 1× PBS.
2. Add 100 μL 0.25 M trypsin to each well, swirl, and incubate in tissue culture incubator for 3–4 min until cells lift off by gently tapping plate. Confirm by examination under a light microscope.
3. Add 400 μL DMEM + 10% HI-FBS to quench the trypsin. Pipette up and down vigorously to remove all cells from well and transfer all 500 μL to 5 mL polystyrene tubes for LSRII/FACS analysis. If necessary, remove any clumps by filtering through a 70 μm filter to prevent clogging the machine.
4. Prepare 10× CCF2/4 master mix as written in Subheading 2.4 and scale as necessary (*see* **Notes 13** and **14**).
5. Add 50 μL 10× CCF2/4 master mix to each FACS tube, incubate at RT in the dark for 20–40 min. Do not allow CCF2/4 incubation to continue for >1.5 h (*see* **Note 14**). Store on ice until FACS analysis.
6. Run samples on FACS machine to determine populations of Yop-TEM$^+$ and Yop-TEM$^-$ cells (*see* Subheading 3.6).

#### 3.3.3 Preparation of Samples for Analysis by Microscopy

1. Remove media from each well. Wash once with 1 mL of 1× PBS.
2. Add 450 μL of PBS to each well and 50 μL of CCF2/4 master mix.
3. Incubate at RT in the dark for 20–40 min.
4. Using Coumarin/Pacific Blue Longpass Filters (Chroma), view on a microscope with 10× and 40× objectives (*see* **Note 15**).

### 3.4 Ex Vivo Infection with Splenocytes or Lung Cells

#### 3.4.1 Harvesting and Infection of Splenocytes or Lung Cells with Yersinia

1. Day 1 afternoon: Prepare overnight cultures of *Yersinia* for infection: Inoculate 2 mL 2× YT plus antibiotics with the *Yersinia* Yop-TEM translational fusion strain. Incubate overnight at 26 °C with aeration.
2. Day 2: Dilute overnight *Yersinia* cultures 1:40 in 2 mL low $Ca^{2+}$ 2× YT (*see* **Notes 8** and **9**).
3. Grow *Yersinia* for 2 h shaking at 26 °C, shift to 2 h shaking at 37 °C.
4. Meanwhile, aseptically harvest spleens or lungs into 1 mL PBS in 2 mL Eppendorf tubes. Keep on ice until all samples are collected (*see* **Note 16**).
5. Prepare single cell suspension by passing organs through 70-μm cell strainer using a syringe plunger into well of 6-well plate. Add 5 mL RPMI to wash cell strainer filter. Mix gently within each well. Transfer to 15 mL conical tubes, add Collagenase D (final concentration 1mg/ml), and incubate for 1 h at 37 °C (*see* **Note 17**).
6. Centrifuge single cell suspensions at 250 × *g* for 5 min at 4 °C.
7. Treat with 5 mL 1× Pharm Lyse for 5 min at 4 °C.
8. Centrifuge at 250 × *g* for 5 min at 4 °C. Remove supernatant.
9. Resuspend in 10 mL RPMI supplemented with 5% HI-FBS. If there are cell clumps, passage a second time through a 70-μm cell strainer into a 50 mL conical tube, and centrifuge at 250 × *g* for 5 min at 4 °C. Count cells.
10. Resuspend cells in RPMI +5% HI-FBS at ~1–2 × $10^6$ cells/mL.
11. Aliquot 200 μL of the cell suspension in an appropriate number of wells in a 24-well plate depending on the number of experimental samples and controls needed. Typically, there are 2–3 replicates per experimental sample and controls for uninfected, CCF2/4-AM staining, and single antibody staining (*see* Table 1 for an example of plate set-up).
12. Take the $OD_{600}$ of diluted bacterial cultures to estimate culture (CFU) concentration, dilute in RPMI to desired concentration.
13. Plate serial dilutions of *Yersinia* used in infection on LB-Irgasan plates to determine MOI. Grow at 26 °C for 2 days before counting CFUs.
14. Add *Yersinia* Yop-TEM translational fusion strain to wells at the desired MOI.
15. Spin for 3 min at 20 × *g* at RT.
16. Incubate at 37 °C for 1–2 h.

**Table 1**
**An example of necessary controls and experimental samples for FACS**

| | 1 | 2 | 3 | 4 | 5 | 6 |
|---|---|---|---|---|---|---|
| FACS gating controls | Uninfected cells unstained | Uninfected cells + antibody 1 | Uninfected cells + antibody 2 | Uninfected cells + antibody 3 | Uninfected cells + antibody 4 | |
| FACS gating controls | | Uninfected cells + CCF2/4 | Infected cells WT ETEM + CCF2/4 | | | |
| Experimental samples | **Uninfected cells + CCF2/ 4 + Antibody combination 1 + 2** | **Uninfected cells + CCF2/ 4 + Antibody combination 3 + 4** | | | | |
| Experimental samples | **Infected cells with WT ETEM + CCF2/ 4+ antibody combination 1 + 2** | **Infected cells with WT ETEM + CCF2/ 4 + antibody combination 3 + 4** | **Infected cells *Yersinia* mutant 1 + CCF2/ 4 + Antibody combination 1 + 2** | **Infected cells *Yersinia* mutant 1 + CCF2/ 4 + Antibody combination 3 + 4** | **Infected cells *Yersinia* mutant 2 + CCF2/ 4 + Antibody combination 1 + 2** | **Infected cells *Yersinia* mutant 2 + CCF2/ 4 + Antibody combination 3 + 4** |

Wells in bold should be infected and stained in either duplicate or triplicate

#### 3.4.2 Preparation of Samples for FACS Analysis

1. After infection, transfer cells by gently pipetting into an Eppendorf tube and centrifuge at 250 × *g* for 5 min at 4 °C.
2. For FACS staining, depending on the number of antibody combinations to be used, resuspend in FACS buffer +100 μg/mL gentamicin at a concentration of 0.5–2 × $10^6$ cells/mL. Remove any clumps of cells by passing through a 70-μM cell strainer or physically with a pipet tip so as to not to clog the FACS machine.
3. Transfer 100 μL of each cell suspension to the appropriate number of 5 mL polypropylene (MoFlo/FACS sorter) or polystyrene (LSRII/FACS analysis) tubes depending on the number of staining combinations to be used. Typically, 50,000 cells are stained per antibody combination, but as few as 1 × $10^4$ or as many as 2 × $10^5$ may need to be stained (*see* **Note 18**). **All of the following steps with CCF2/4-AM and antibody staining must be kept in the dark as much as possible.** This includes keeping FACS samples in foil until it is time to FACS them.
4. Incubate cells with 10 μL 10× CCF2/4-AM mixture without gentamicin per 100 μL of cells for 15 min at room temperature (22–24 °C). Make sure to leave aside sufficient aliquots of cells that are not treated with CCF2/4-AM to serve as gating controls for antibodies (*see* **Note 19** and Table 1).
5. Add 50 μL 1× Fc block per 100 μL of cells. Incubate 10 min at 4 °C.
6. Aliquot 50 μL of desired antibody combinations into each tube (*see* **Notes 20** and **21**).
7. Incubate for 15–30 min at 4 °C. If in a 96-well plate format, transfer to appropriate FACS tubes.
8. Add 200 μL of PBS into each tube and analyze immediately by FACS (Subheading 3.6).

### 3.5 Mouse Infection

#### 3.5.1 Intranasal Infection of Mice

1. Day 1 afternoon: From either a glycerol stock or a colony on a fresh plate, inoculate 2 mL of 2× YT with *Yersinia*. Grow overnight at 26 °C with aeration.
2. Day 2 AM: Dilute overnight cultures 1:40 in 2 mL fresh 2× YT and allow cultures to grow for 8 h at 26 °C with aeration.
3. Day 2 PM: Dilute 8 h cultures 1:100 into 2 mL fresh 2× YT at 26 °C and grow overnight with aeration (*see* **Note 22**).
4. Day 3: Take $OD_{600}$ of overnight bacterial cultures to estimate culture concentration (CFU/mL), and dilute to appropriate CFU in sterile 1× PBS. We normally infect mice with 500 CFU of wild-type *Yersinia pseudotuberculosis* IP2666 intranasally per mouse (*see* **Note 23**). Plate serial dilutions of *Yersinia* used in

infection on LB-Irgasan plates to determine MOI. Grow at 26 °C for 2 days before counting CFUs.

5. Administer anesthetics prior to intranasal infection (*see* **Note 24**). For isoflurane, set up anesthesia chamber with $O_2$ at 1–2.5 psi and the isoflurane at 2.5–3 psi.
6. Place mice in chamber until their breathing slows and they remain completely limp upon stimulus such as gentle rocking of the chamber.
7. Remove mouse from isoflurane chamber. Holding the mouse at the scruff of the neck its with nose pointed up, pipet 40–50 μL of bacteria in PBS slowly onto the nose drop by drop (*see* **Note 25**). Make sure that the droplets are placed directly on the nostril openings. The mouse should inhale these droplets quickly. The speed of dropping the droplets should match that of the mouse's inhalation and breathing.
8. Once the entire inoculum is inhaled, gently place the mouse in cage on area cleared of bedding. Monitor the mouse until it regains consciousness, and make sure it does not immediately cough up the bacteria.
9. Monitor the mice once or twice daily over the course of the infection.

#### *3.5.2 Post-infection Organ Harvest*

1. Collect organs in 1 mL PBS in 2 mL Eppendorf tubes. Keep on ice until the single cell suspension is prepared (*see* **Note 16** for spleens during systemic infection following intranasal or intravenous injection).
2. Prepare the single cell suspension by passing organs through a 70-μm cell strainer using a syringe stopper into a well of a 6-well plate. Add 5 mL RPMI to the cell strainer and flush into the well. Mix thoroughly.
3. Transfer 100 μL to an Eppendorf tube for plating for CFU (*see* **Note 26**) and the rest to a 15 mL conical tube containing a final concentration of 1 mg/mL collagenase D and 100 μg/mL gentamicin.
4. Incubate at 37 °C for 30 min. Invert tubes 4–6 times (*see* **Note 17**).
5. Incubate at 37 °C for another 30 min. At this point, cells should appear dispersed within the suspension (*see* **Note 27**).
6. Centrifuge at 250 × *g* at 4 °C for 5 min. Remove supernatant. Resuspend the pellet in 5 mL 1× Pharm Lyse. Incubate on ice for 5 min.
7. Centrifuge at 250 × *g* at 4 °C for 5 min. Remove supernatant.
8. Continue with **step 11** of Subheading 3.4.2 for loading cells with the CCF2/4-AM mixture and FACS staining with antibodies to surface markers.

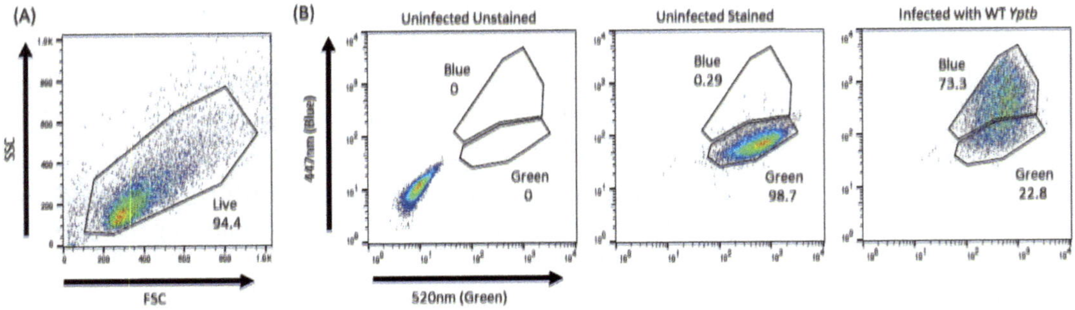


**Fig. 1** Gating strategy for FACS. (**a**) Representative image of using FSC vs. SSC plot to gate on live cells. (**b**) HT29 cells were infected with WT IP2666 expressing YopE-TEM at an MOI of 10:1 for 1 h or left uninfected. After infection, cells were harvested and stained with CCF4-AM for 20 min or left unstained. Samples were then analyzed by flow cytometry to determine green (520 nm) and blue (447 nm) cell populations, where green represents no translocation and blue represents translocation of the Yop-TEM. Left: Uninfected unstained samples are negative for both green and blue fluorescence. Middle: Uninfected stained samples are positive for green fluorescence but negative for blue fluorescence. Right: Infected stained samples are positive for both green and blue fluorescence

### *3.6 Analytical FACS Analysis*

1. Keep all samples in the dark and on ice until they are ready to be analyzed by FACS.
2. Set up the FACS machine using the 409 nm laser and detection filters for 447 nm and 520 nm emission. Prepare to collect forward scatter control (FSC) vs. side scatter control (SSC), 447 nm vs. 520 nm, as well as all other fluor channels needed for antibody detection (*see* **Note 28**).
3. First, run the unstained (non-CCF2/4-treated and no antibodies stained) control to set the voltages of the lasers (*see* **Notes 29–31**, Table 1 and Fig. 1a).
4. Run the CCF2/4-AM-treated uninfected control without antibody. This sample will define the "green-positive" or TEM$^-$ cells on the 520 nm axis as those cells that have shifted from the unstained sample (*see* Fig. 1b).
5. Run the CCF2/4-AM-treated control cells infected with Yop-TEM expressing *Yersinia* but not treated with any antibodies (*see* **Note 11**). If Yop-TEM translocation occurred, there will be a shift of cells toward the 447 nm axis, which are blue, or TEM$^+$ cells. Cells that remain positive on only the 520 nm axis are TEM$^-$, which are green (*see* Fig. 1b).
6. If staining cells for surface markers using antibodies, run single stained controls (*see* **Note 20c** for consideration of tandem-conjugated antibodies). Antibodies using fluorophores that are excited by the same laser may need compensation.
7. Run all additional samples using the voltages and channels set in **steps 3–6**.
8. After sample collection, export the data as FCS files.

9. Using a FACS analysis program of choice, analyze the data. *See* Fig. 1 for an example of blue/green gating strategy.

### 3.7 Using Yop-TEM+ and Yop-TEM− Cells for Functional and Analytical Assays

1. Using the same staining (*see* Subheading 3.4.2) and gating (*see* Subheading 3.6 and Fig. 1) strategy, sort Yop-TEM+ and Yop-TEM− cells into sorting buffer in 15 mL conical tubes (*see* Fig. 1). Store on ice until all samples are sorted (*see* **Note 32**).
2. Sorted Yop-TEM+ and Yop-TEM− cells can be used in the following assays.

#### 3.7.1 Preparation of PMNs for Functional Assays

1. For functional assays with PMNs, rest FACS-sorted cells at room temperature in sorting buffer for 1 h.
2. Centrifuge at 250 × *g*, remove supernatant, and resuspend in 5 mL 1× $Ca^{2+}$- and $Mg^{2+}$-free PBS, or HBSS. Rest 30 min at room temperature.
3. Centrifuge at 250 × *g*, remove supernatant and resuspend in 1× PBS or HBSS with $Ca^{2+}$, $Mg^{2+}$, and 5 mM glucose at ~$1 \times 10^6$ cells/mL. Rest 30 min at room temperature.
4. Shift cell suspension to 37 °C for 15 min.
5. PMNs are now ready for functional assays such as production of reactive oxygen species.

#### 3.7.2 Preparation of Cells for Western Blotting

1. Spin down FACS-sorted cells and lyse in SDS sample buffer for a final cell concentration of $0.25–1 \times 10^8$ cells/mL (*see* **Note 33**).
2. Boil samples at 95 °C for 5 min. Run samples onto gel immediately or store at −20 °C until analysis.
3. Load $2.5 \times 10^5–1 \times 10^6$ cell equivalents on a SDS-PAGE gel– gel percentage is dependent on proteins of interest or use a 4–20% gradient gel.
4. Use transfer and western blotting protocols of your choice.

#### 3.7.3 Preparation of Cells for qRT-PCR

1. Spin down FACS sorted cells, and lyse in Trizol at ~$5–10 \times 10^6$ cells/mL per manufacturer's instructions (*see* **Notes 33** and **34**). Samples can be stored at −80 °C until use.
2. Continue with RNA extraction using desired procedure (e.g., QIAGEN RNeasy Mini Kit).

#### 3.7.4 Preparation of Cells for Cytokine Analysis

1. For in vivo mouse infections, harvest infected tissue in the presence of BD GolgiPlug according to manufacturer's instructions, or 3 μg/mL Brefeldin A (BFA). **All steps following tissue harvest must be in media containing either GolgiPlug or 3 μg/mL BFA** (*see* **Note 35**).
2. Spin down FACS-sorted cells and resuspend in RPMI +10% FBS + 3 μg/mL BFA at a concentration of ~$5 \times 10^5–1 \times 10^6$ cells/mL (*see* **Note 36**).

3. Follow manufacturer's instructions for BD Cytometric Bead Array (*see* **Note 37**).
4. Alternatively, follow an intracellular staining protocol of choice. Most antibody manufacturers have an established protocol optimized for their reagents (*see* **Note 38** for brief intracellular staining protocol, and **Note 39** for consideration of cytokine antibodies).
5. Using gating strategy from Subheading 3.6, analyze by FACS.

### 3.8 High-Throughput Analysis by Plate Reader

1. Day 1: Seed cultured cells in a 384-well plate: $1 \times 10^4$ cells/well in 25 μL RPMI with 5% HI-FBS or DMEM with 10% HI-FBS.
2. Day 1 afternoon: Prepare overnight cultures of *Yersinia* for infection: Inoculate 2 mL 2× YT plus antibiotics with *Yersinia* Yop-TEM translational fusion strain. Incubate overnight at 26 °C with aeration.
3. Day 2: Dilute overnight *Yersinia* cultures 1:40 in 2 mL low $Ca^{2+}$ 2× YT (*see* **Notes 8** and **9**).
4. Grow *Yersinia* for 2 h shaking at 26 °C, and then shift to 2 h shaking at 37 °C.
5. Take $OD_{600}$ of cells to estimate culture concentration, dilute in tissue culture media to desired concentration in 2 μL.
6. Plate serial dilutions of *Yersinia* used in infection on LB-Irgasan plates to determine MOI. Grow at 26 °C for 2 days before counting CFUs.
7. Using a robotic machine, add 3 μL of the 10× CCF2-AM mixture to each well of 384-well plates (*see* **Note 40**).
8. Incubate the plate at 30 °C for 30 min.
9. Add small molecule compound libraries with robotics, if desired.
10. Infect with *Yersinia* at the desired MOI. Include at least 8–12 wells/plate with each of the following controls: cells with CCF2/4-AM, cells alone, and CCF2/4-AM alone to control for background fluorescence signal. Use WT *Yersinia* expressing Yop-TEM as a positive control and a T3SS-defective *Yersinia* as a negative control on each plate.
11. Spin plate at $20 \times g$ for 3 min.
12. Place the plate in a tissue culture incubator and allow infection to proceed for 1–2 h.
13. To stop infection, add 100 μg/mL gentamicin to each well.
14. Use a plate reader to measure blue fluorescence (447 nm) before green fluorescence (520 nm) (*see* **Note 41**).

## 4 Notes

1. Any compatible bacterial allelic exchange vector containing kanamycin resistance and a SacB gene can be used, for example, pRS47 or derivatives of pCVD442 that have replaced the ampicillin resistance cassette with another drug resistance cassette [25]. Alternatively, Yop-TEM fusion constructs can be generated on replication-competent vectors, such as the pMMB series of vectors that are selectable with antibiotics other than ampicillin [9]. If using a replication-competent plasmid, the plasmid can be introduced by electroporation.
2. For amplification of TEM-1 from the *blaM* gene from pBR322 with primers containing restriction sites 5′ NotI; 3′ SacI [25]:

   TEM1F 5′ GAGAGAGCGGCCGCCACCCAGAAACGC TGGTG 3′.

   TEM1R 5′ AGACAGAGCTCGCATGCTGAGTAAACTT GGTCTGACAGT 3′.
3. To inactivate the complement system, and generate heat-inactivated (HI)-FBS, incubate thawed FBS at 55 °C for 30 min. Complement factors can kill *Yersinia,* which can affect the multiplicity of infection (MOI) on the cells and skew CFU plating.
4. Instead of a native *yop* promoter, an inducible promoter such as IPTG, arabinose, or Tet promoter can be used if expression needs to be uncoupled from control of T3SS system.
5. Ensure that the Yop coding sequence is in frame with the TEM and that there are no stop codons between the two open reading frames.
6. The gene encoding the Yop-TEM construct can be inserted at a neutral site in the virulence plasmid or chromosome if desired.
7. *Yersinia* should be grown at 26 °C; *E. coli* can be grown at 37 °C, but conjugation is performed at RT or 26 °C overnight. When growing *Yersinia* in liquid culture, Irgasan should not be added to media. Irgasan is only added to agar plates.
8. Low $Ca^{2+}$ 2× YT should always be made fresh from filter sterilized or autoclaved components.
9. The $OD_{600}$ of overnight cultures should be measured to normalize the ODs before doing the 1:40 dilutions to best ensure comparable numbers of bacteria after 4 h.
10. For adherent, epithelial-like cell lines such as Hela, HEp-2, Caco-2, and HT-29 cells, 0.8–1.0 × $10^5$ cells/well/24-well plate are added in 1 mL DMEM with 10% heat-inactivated (HI)-FBS. For adherent, macrophage cell lines, such as RAW264.7 or J774, 2 × $10^5$ cells/well/24-well plate can be

seeded in 1 mL DMEM with 10% heat-inactivated (HI)-FBS. Depending on the growth rate of each cell line, adherent cells should be at ~80–90% confluency on the day of *Yersinia* infection. For adherent cells, adjust concentration based on surface area of 6, 12, 48, 96, or 384-well plates. For suspension cells, such as Jurkat cells and Raji B, cell concentrations can range from $2 \times 10^5$ to $1 \times 10^6$ cells/mL.

11. Leave at least two wells that are not infected with bacteria as negative controls. Add CCF2/4 to one of these wells. Also, include WT *Yersinia* without Yop-TEM construct or a T3SS mutant expressing Yop-TEM as a negative control for translocation.
12. Epithelial cells, such as Hela and HEp-2 cells infected with T3SS-competent *Yersinia* expressing YopE should begin to round up after about 1 h when infected at an MOI of 10:1 or greater, which will be visible under light microscopy. This is not apparent for most immune cells due to the fact that they are usually round.
13. All of the following CCF2/4-AM and antibody staining steps need to be kept in the dark as much as possible. It is also critical that the CCF2/4 mixture be prepared fresh for each experiment and that it be kept in the dark. This includes keeping the samples that will be run on the FACS machine wrapped in foil until time to run the samples.
14. CCF2 vs. CCF4: CCF4 was engineered to have better solubility in aqueous solutions; however, our unpublished observations indicate that CCF4 is more rapidly turned over within cells compared to CCF2. Thus for applications where cells are loaded prior to infection [25], CCF2-AM is a better substrate because it lasts longer. In ex vivo and in vivo tissue infections, either CCF2-AM or CCF4-AM can be used as long as loading of cells with CCF2/4-AM occurs after infection and translocation of Yop-TEM.
15. This signal quenches within 30 s. Be quick!
16. To liberate all cells in spleens, including dendritic cells from spleen samples, perfuse tissues with 1 mg/mL Collagenase D for 30 min at 37 °C prior to passing through the filter. To perfuse spleens, inject several areas of the spleen with Collagenase D using a 30-gauge needle and syringe using about 50 μL of Collagenase D solution. Submerge the spleen in a small volume of the solution.
17. To help liberate cells from lungs, invert tubes 2–3 times every 15–20 min.
18. Alternatively, all the staining steps can be done in 96-well round, U, or V bottom plates. The resuspended cells in

FACS buffer +100 μg/mL gentamicin should be aliquoted to an appropriate number of wells in the plates. Typically, 50,000 cells should be stained per condition. Do not add more than 250 μL of the suspension to each well in the 96-well plate as larger volumes will contaminate adjacent samples (*see* Table 1).

19. At least one unstained (no antibodies/no CCF2/4-AM added) control and single antibody and CCF2/4-AM-only control samples are required. Also, set aside an aliquot from uninfected and infected CCF2/4-AM-treated cells to serve as gating controls (*see* Table 1).

20. During the bacterial infection, prepare antibody combinations in FACS buffer. As the required dilution of the antibody can differ between manufacturer and lot, each lot of antibody should be titrated with appropriate tissue sample prior to use with CCF2/4-AM to optimize signal-to-noise ratio reduce spillover between fluorophores.
    (a) To titrate the antibody, aseptically obtain cells (i.e. from the spleens or lungs) of a mouse. Prepare a single cell suspension and lyse RBCs (*see* Subheading 3.5). Leave one unstained sample for gating control. Add a single antibody at 1:50, 1:100, 1:250, and 1:500 dilution in FACS buffer to the cell suspension.
    (b) The emission spectra of fluorescently conjugated antibodies should not overlap with those of CCF2/4-AM, namely 447 and 520 nm. Good choices are PE, PE-Cy5, PE-Cy7, and APC, but not FITC and PacBlue.
    (c) The tandem-conjugated antibodies, such as PE, PE-Cy5, and PE-Cy7, often require greater compensation than APC. Therefore, it is necessary to have single-stained controls to set laser intensity (voltages) and compensation.

21. We have used the following combinations for white blood cells.
    (a) CD45: all WBCs.
    (b) CD11b: all phagocytes.
    (c) $CD11b^+$ $Ly6G^+$: PMNs only.
    (d) $CD11b^{int}$ $CD11c^{hi}$: alveolar macrophages.
    (e) $CD11b^{int}$ $CD11c^{hi}$: dendritic cells.
    (f) $Gr1^{lo}$ $CD11b^+$: resident monocytes.
    (g) $CD4^+$ $TCR\beta^+$: CD4 T cells.
    (h) $CD8^+$ $TCR\beta^+$: CD8 T cells.
    (i) $B220^+$ $CD19^+$: B cells.

22. Preparation of overnight cultures for infection of mice may differ between *Yersinia* strains and laboratories.

23. Required infectious doses will vary greatly depending on the *Yersinia* strain, mouse strain and route of infection. To detect TEM$^+$ cells, overall translocation levels above 2–3% are required because the background CCF2/4$^+$ staining in non-infected cells is around 0.5–1% of cells. Typically, translocation levels of 5–75% are optimal. To achieve these levels, the MOI and/or length of infection may need to be optimized.
24. General anesthetics can be administered to mice either as injectable (i.e. ketamine) or inhaled (i.e. isoflurane). Consideration for choice of anesthetics includes strain, age, weight, disease model, and experimental protocol [27].
25. Intranasal delivery of bacteria requires 40–50 μL of PBS to ensure that the inoculum will reach the lungs. A lower volume will more likely result in the infection of the nasopharynx. Make sure that the mouse breathes in all the inoculum with no bubbling or pooling in the mouth. Success of intranasal delivery varies between mouse strains. For mouse strains like C57B/6, where there is observed increased tendency for bubbling and pooling, intratracheal delivery should be used with isoflurane. For BALB/c, CD-1, and Swiss-Webster mice, intranasal infection works well with isoflurane.
26. A portion of the cell suspension can be diluted 1:1 in 50% glycerol and frozen for future repeated plating of CFU. Additionally, the cell suspension can be spun down at 17,000 × *g*, and the supernatant can be used for ELISA assays or cells can be used for qRT-PCR.
27. If the cells are aggregating toward the bottom of the tube or appear to be clumping when tube is inverted, pellet the cell suspension at 250 × *g* for 5 min, and resuspend the pellet in fresh RPMI with 1 mg/mL Collagenase D + 100 μg/mL gentamicin for an additional hour at 37 °C.
28. Both FACS and plate reader machines should be equipped with a 409 nm laser for excitation, as well as 447 and 520 nm filters for emission.
29. Make sure the unstained cells are visible on both the forward scatter control (FSC) and side scatter control (SSC). FSC measures the size of the cells, while SSC measures the granularity (internal complexity) of the cells.
30. The voltages of the lasers (i.e. 409 nm laser for CCF2/4) that will be used to analyze the FRET should be set to establish a negative population with the unstained control sample. Most cells emit a low level of fluorescence when excited ("autofluorescence"), and, therefore, will show up on the plot of 447 nm vs. 520 nm and relevant antibody fluors. This serves as the "negative" population for these channels. For all subsequent samples, cells that appear on the plot above this

baseline "negative" population will be considered as positive cells (*see* Fig. 1).

31. Each setup and gating strategy is subjective to the person running the samples. It is critical that these settings are not changed between samples within the same experiments. They may be different between experiments. The investigator must note all settings to compare data analysis appropriately across multiple experiments.
32. The rate of sorting will depend on the concentration of cells in each tube and FACS sorter. If a tube or series of tubes takes greater than 30 min, it is optimal to stagger your loading with CCF2/4 and staining of surface antibodies in order to reduce CCF2/4 turnover.
33. The number of Yop-TEM$^+$ and Yop-TEM$^-$ cells can vary between mouse, organs, cell populations of interest, and strains used for infection. Thus, pooling cells from more than one mouse is typically required so that at least $5 \times 10^5$ will be analyzed by western blot. A total of approximately $1–2 \times 10^6$ Yop-TEM$^+$ and Yop-TEM$^-$ can be collected per wild-type *Yersinia*-infected mouse lung made up of 20–70% neutrophils, 1–4% alveolar macrophages, ~5% resident monocytes, ~5% dendritic cells, 2–10% CD8 T cells, 5–20% CD4 T cells, and 5–15% B cells.
34. Different cell types will have different RNA quantities and extraction efficiency; for example, efficiency of RNA extraction from neutrophils is significantly lower than other cells found in lungs or spleens. Ideally, 1 μg of RNA/condition is used as template for qRT-PCR, but smaller amounts can be used if required. Frequently, we have found that we can use TEM$^+$ and TEM$^-$, PMNs, and macrophages from one mouse lung or spleen for qRT-PCR analysis, in contrast to the number of mice required for Western Blot analysis and cytokine analysis.
35. Ex vivo addition of either GolgiPlug or 3 μg/mL BFA to cells blocks their intracellular protein transport processes preventing cytokines from being secreted into the supernatant during tissue processing. Optimal BFA concentration is dependent on cell types and may need titration depending on manufacturer's recommendation. Too low will not sufficiently block protein transport, and too high will cause cells to die during culture.
36. Pooling cells between multiple mice might be necessary. Generally, for intracellular staining, 50,000 cells are stained per antibody combination, but as few as $1 \times 10^4$ or as many as $2 \times 10^5$ may need to be stained. For intracellular staining, cells may need to be incubated and/or stimulated ex vivo for 1–6 h (depending on cell types) for accumulation of the cytokines to

reach a certain concentration (depending on the antibody and the cytokine) allowing for detection by flow cytometry.

37. For the BD Cytometric Bead Array, it might be necessary to dilute test samples of known concentration or assumed to contain high levels of a given cytokine by a desired dilution factor. For each test sample, we recommend making a couple of dilutions (i.e. 1:2, 1:10, or 1:50) to ensure that the median fluorescence values will fall within the generated standard curve.
38. Typical intracellular staining protocols for sorted cells may include (1) an initial wash with FACS buffer, (2) fixation with 1–4% formaldehyde (PFA) at room temperature for 10–20 min, (3) two washes with 0.1% saponin in FACS buffer for permeabilization, (4) intracellular staining with antibodies for cytokines in 0.1% saponin in FACS buffer, (5) a final wash with 1× PBS, and (6) suspension in 1× PBS for FACS analysis.
39. When choosing antibodies for cytokine analysis, avoid fluorophores that have excitations or emissions overlapping with CCF2 staining (excitation: 409 nm/emission: 447 nm and 520 nm), or any previously used surface antibodies. There will be residual CCF2 staining post-fixation, and antibodies to surface markers will still be detected.
40. CCF2-AM is preferred because it has a longer half-life and must be added before infection occurs. To ensure sterility of the robotics machine and prevent contamination with a BSL2 pathogen, all compounds are added before *Yersinia*.
41. To determine the values for green and blue fluorescence in each well, please see the Methods of Harmon DE, Davis AJ, Castillo C, et al. (2010) Identification and characterization of small-molecule inhibitors of Yop translocation in *Yersinia pseudotuberculosis*. Antimicrob Agents Chemother 54:3241–54 [25].

## Acknowledgments

We thank all previous Mecsas laboratory members who used and optimized this technique in various procedures for their insights and notes. This work was supported by NIH AI R01 AI113166 and R01 AI107055.

## References

1. Gemski P, Lazere JR, Casey T et al (1980) Presence of a virulence-associated plasmid in *Yersinia pseudotuberculosis*. Infect Immun 28:1044–1047
2. Portnoy DA, Falkow S (1981) Virulence-associated plasmids from *Yersinia enterocolitica* and *Yersinia pestis*. J Bacteriol 148:877–883
3. Cornelis GR, Boland A, Boyd AP et al (1998) The virulence plasmid of *Yersinia*, an antihost

genome. Microbiol Mol Biol Rev 62:1315–1352

4. Dewoody RS, Merritt PM, Marketon MM (2013) Regulation of the *Yersinia* type III secretion system: traffic control. Front Cell Infect Microbiol 3:4
5. Chung LK, Bliska JB (2016) *Yersinia* versus host immunity: how a pathogen evades or triggers a protective response. Curr Opin Microbiol 29:56–62
6. Pinaud L, Sansonetti PJ, Phalipon A (2018) Host cell targeting by enteropathogenic bacteria T3SS effectors. Trends Microbiol 26 (4):266–283
7. Charpentier X, Oswald E (2004) Identification of the secretion and translocation domain of the enteropathogenic and enterohemorrhagic *Escherichia coli* effector cif, using TEM-1 beta-lactamase as a new fluorescence-based reporter. J Bacteriol 186:5486–5495
8. Zlokarnik G, Negulescu PA, Knapp TE et al (1998) Quantitation of transcription and clonal selection of single living cells with beta-lactamase as reporter. Science 279:84–88
9. Marketon MM, DePaolo RW, DeBord KL et al (2005) Plague bacteria target immune cells during infection. Science 309:1739–1741
10. Durand EA, Maldonado-Arocho FJ, Castillo C et al (2010) The presence of professional phagocytes dictates the number of host cells targeted for Yop translocation during infection. Cell Microbiol 12:1064–1082
11. Köberle M, Klein-Günther A, Schütz M et al (2009) *Yersinia enterocolitica* targets cells of the innate and adaptive immune system by injection of Yops in a mouse infection model. PLoS Pathog 5:e1000551
12. Sory MP, Cornelis GR (1994) Translocation of a hybrid YopE-adenylate cyclase from *Yersinia enterocolitica* into HeLa cells. Mol Microbiol 14:583–594
13. Sutcliffe JG (1978) Nucleotide sequence of the ampicillin resistance gene of *Escherichia coli* plasmid pBR322. Proc Natl Acad Sci U S A 75:3737–3741
14. O'Callaghan CH, Morris A, Kirby SM et al (1972) Novel method for detection of beta-lactamases by using a chromogenic cephalosporin substrate. Antimicrob Agents Chemother 1:283–288
15. Maldonado-Arocho FJ, Green C, Fisher ML et al (2013) Adhesins and host serum factors drive yop translocation by *Yersinia* into professional phagocytes during animal infection. PLoS Pathog 9:e1003415
16. Pechous RD, Sivaraman V, Price PA et al (2013) Early host cell targets of *Yersinia pestis* during primary pneumonic plague. PLoS Pathog 9:e1003679
17. Paczosa MK, Fisher ML, Maldonado-Arocho FJ et al (2014) *Yersinia pseudotuberculosis* uses Ail and YadA to circumvent neutrophils by directing Yop translocation during lung infection. Cell Microbiol 16:247–268
18. Rolán HG, Durand EA, Mecsas J (2013) Identifying *Yersinia* YopH-targeted signal transduction pathways that impair neutrophil responses during in vivo murine infection. Cell Host Microbe 14:306–317
19. Dewoody R, Merritt PM, Marketon MM (2013) YopK controls both rate and fidelity of Yop translocation. Mol Microbiol 87:301–317
20. Keller B, Mühlenkamp M, Deuschle E et al (2015) *Yersinia enterocolitica* exploits different pathways to accomplish adhesion and toxin injection into host cells. Cell Microbiol 17:1179–1204
21. Autenrieth SE, Linzer T-R, Hiller C et al (2010) Immune evasion by *Yersinia enterocolitica*: differential targeting of dendritic cell subpopulations in vivo. PLoS Pathog 6:e1001212
22. Zhang Y, Tam JW, Mena P et al (2015) CCR2+ inflammatory dendritic cells and translocation of antigen by type III secretion are required for the exceptionally large CD8+ T cell response to the protective YopE69-77 epitope during *Yersinia* infection. PLoS Pathog 11:e1005167
23. Houppert AS, Kwiatkowski E, Glass EM et al (2012) Identification of chromosomal genes in *Yersinia pestis* that influence type III secretion and delivery of Yops into target cells. PLoS One 7:e34039
24. Dewoody R, Merritt PM, Houppert AS et al (2011) YopK regulates the *Yersinia pestis* type III secretion system from within host cells. Mol Microbiol 79:1445–1461
25. Harmon DE, Davis AJ, Castillo C et al (2010) Identification and characterization of small-molecule inhibitors of Yop translocation in *Yersinia pseudotuberculosis*. Antimicrob Agents Chemother 54:3241–3254
26. Duncan MC, Wong WR, Dupzyk AJ et al (2014) An NF-κB-based high-throughput screen identifies piericidins as inhibitors of the *Yersinia pseudotuberculosis* Type III secretion system. Antimicrob Agents Chemother 58:1118–1126
27. Gargiulo S, Greco A, Gramanzini M et al (2012) Mice anesthesia, analgesia, and care, Part I: anesthetic considerations in preclinical research. ILAR J 53:E55–E69

# Chapter 10

# Monitoring of Neutrophil Recruitment to Mice Lungs During Pneumonic Plague

**Yaron Vagima, Yinon Levy, and Emanuelle Mamroud**

## Abstract

Early sensing of bacterial infection and the immediate recruitment of neutrophils to the lung is a major and decisive stage of the innate immune response to pulmonary bacterial infections. This chapter details the preparation of lung tissue suspensions from mice infected intra-nasally (I.N.) with the plague bacterium *Yersinia pestis* to study *in vivo* neutrophil responses to the infection. The samples were used for the quantification of neutrophil levels and for the characterization of the pro-inflammatory response required for neutrophil recruitment to the lung. The specific requirements for performing the procedures under Biosafety Level 3 containment and the proper handling and sterilization of the samples are discussed.

**Key words** Plague, Infection, *Y. pestis*, Neutrophils isolation, Lung samples, Mouse model, BSL-3, FACS analysis

## 1 Introduction

The immediate recruitment of neutrophils to the lung during pulmonary bacterial infection is a major and critical part of the innate immune response. Early after infection, neutrophils are mobilized from the bone marrow to the blood within a few hours, and they then migrate to the infected site along with a chemotactic gradient of specific chemoattractants, including CXCL1 and CXCL2, produced by cells that reside in the inflamed lung [1, 2]. To allow circulating neutrophils to cross the vasculature and to arrive at the site of infection, several adhesion molecules are expressed on endothelial cells adjacent to the inflamed tissue. E- and P-selectins are involved in the initial attachment and rolling of the neutrophils on the endothelium. Vascular cell adhesion molecule 1 (VCAM-1) and intracellular adhesion molecule 1 (ICAM-1) mediate the subsequent steps, in which the neutrophils tightly adhere to the endothelium and transmigrate to the inflamed tissue [3]. After entering the lungs, neutrophils phagocytose and digest the invading pathogen and release a variety of bactericidal

Viveka Vadyvaloo and Matthew B. Lawrenz (eds.), *Pathogenic Yersinia: Methods and Protocols*, Methods in Molecular Biology, vol. 2010, https://doi.org/10.1007/978-1-4939-9541-7_10, © Springer Science+Business Media, LLC, part of Springer Nature 2019

molecules and pro-inflammatory cytokines that modulate the host defense response to the infection [4].

The Gram-negative pathogen *Y. pestis* is the causative agent of plague and has been responsible for pandemics resulting in the death of millions worldwide. The possible use of *Y. pestis* as a bio-terrorism agent threatens public health and has led to the classification of this pathogen as a Tier 1 Select Agent by the CDC (http://www.cdc.gov). Inhalation of droplets or aerosols containing the pathogen leads to the development of primary pneumonic plague, a rapidly progressive fatal disease that can spread from person to person [5–7].

Pneumonic plague has been intensively investigated during the last two decades using animal models. It was characterized as a biphasic disease [8]. In the early stages of post intra-nasal (I.N.) infection, significant bacterial replication was observed in the lungs without an adequate response of the host innate immune system [8–13]. The immune response at the site of infection was delayed, although effective mobilization of both mature and immature cells was observed from the bone marrow, followed by release of neutrophils into the blood stream [14]. It, therefore, appears that the interference of *Y. pestis* with the expression of key neutrophil chemoattractants KC, MIP-2, and G-CSF, as well as key leukocyte adhesion molecules in the lungs [13], prevent the rapid recruitment of neutrophils from circulation into the lung, thus allowing the replication of the pathogen at the infection site [8–14].

The type III secretion system of *Y. pestis* includes a secretion apparatus, chaperones, and several effector proteins called Yops that allow the pathogen to evade the host innate immunity during the early stages of the disease [15, 16]. The Yops are translocated from the pathogen into the host cell cytosol, leading to disruption of the complex signaling networks involved in cytokine and chemokine secretion, phagocytosis and cell survival [16–21].

The delayed immune response rapidly turns into a massive and uncontrolled response, characterized by infiltration of massive amounts of neutrophils into the lung. High levels of MMP8 and MMP9 metalloproteinases are produced and released by the infiltrating neutrophils [14]. At this stage of the infection, the architecture of the lungs is affected, the blood-air barrier is disturbed and bacterial dissemination to internal organs and blood leads to the fatal outcome of the infection [8–14].

In this chapter, we detail the preparation of lung samples from mice infected I.N. with *Y. pestis* (Fig. 1). We used these protocols to study the cascade of neutrophil recruitment from the circulation into the lungs during pneumonic plague and to identify the pathogen's virulence factors involved in suppressing this process [13]. The protocol detailed in this chapter strictly complies with the international guidelines for the use and handling of pathogenic microorganisms (http://www.cdc.gov) and with recommendations as per the Care and Use of Laboratory Animals of the National Institute of Health.

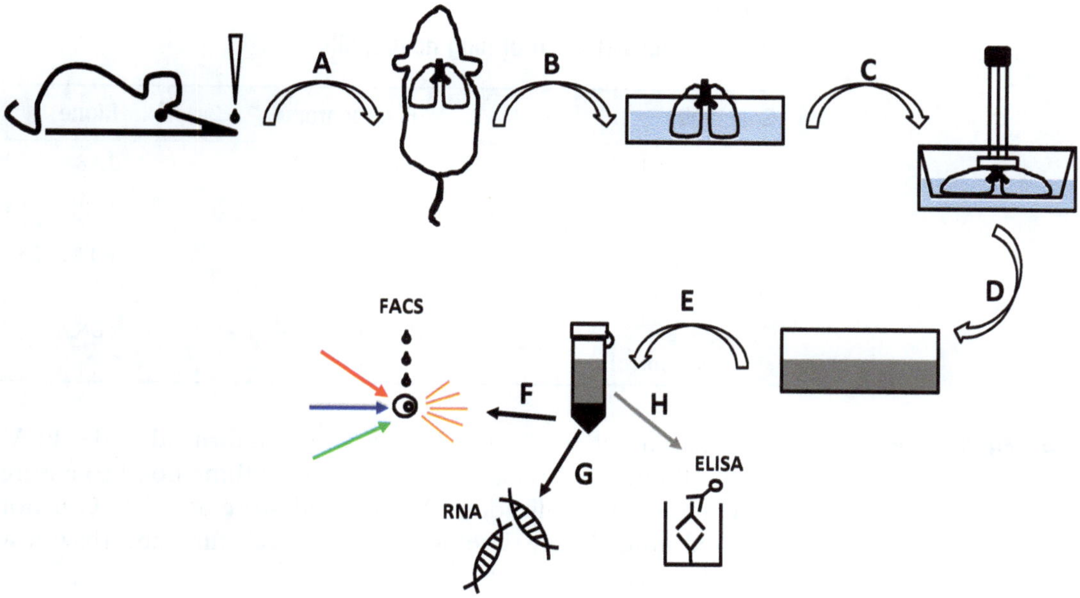


**Fig. 1** A step-by-step flowchart of infected lung extraction and the evaluation of neutrophils, chemoattractants, and RNA levels in the lung. Mice are intra-nasally infected with *Y. pestis*, while at a specific time-point post-exposure, mice are sacrificed (**a**) and the whole lung is transferred to a 6-well plate containing cold PBS (**b**). A lung single-cell suspension is prepared by mashing the organ against a 70 μM cell strainer using a syringe plunger (**c**), thus allowing the lung contents to be released into the cold PBS (**d**). Lung extracts are transferred into a new tube and cells are pelleted by centrifugation (**e**). The total lung cell pellets are further analyzed for immune cell distribution using FACS acquisition and analysis (**f**). The inflammatory response in the lungs was further characterized in lung cells by analyzing RNA expression profiles (**g**) and by chemoattractant quantitation in lung supernatants using ELISA (**h**)

## 2 Materials

### 2.1 Mice

1. Female inbred C57BL/6 mice (6–10 weeks old) (*see* **Note 1**).

### 2.2 Lung Removal and Processing

1. 70% ethanol (EtOH) cleaning solution.
2. 1× PBS.
3. Protease inhibitor cocktail 1% solution in 1× PBS. Prepare fresh solution and keep on ice until use.
4. Forceps and seizures.
5. Sacrifice board.
6. Six-well plate.
7. 70 μm Cell strainer
8. 3 ml Syringe.

### 2.3 Cell Counting

1. Inverted microscope.
2. Hemocytometer and cover slips.
3. Trypan blue solution.

**Table 1**
**Antibodies for detection of lung neutrophils**

| Specificity | Fluorochrome | Clone |
|---|---|---|
| Anti-mouse Ly6G | APC | 1A8 |
| Anti-mouse CD11b | PerCP-Cy5.5 | M1/70 |
| Rat IgG2a (Isotype control) | APC | RTK2738 |
| Rat IgG2a (Isotype control) | PerCP-Cy5.5 | eBR2a |

### 2.4 FACS Analysis

1. Fixation buffer 4% PFA: Dissolve 4 g paraformaldehyde (PFA) in 100 ml 1× PBS and heat to 70 °C in fume hood to ensure complete solubilization. Aliquot and store at −20 °C if not used immediately. Use new solution each time and thaw just before use.
2. Cold 1× PBS.
3. Fluorescence-activated cell sorting (FACS) buffer: 3% fetal calf serum (FCS), 2 mM EDTA in 1× PBS. Store at 4 °C.
4. Antibodies for detection of lung neutrophils obtained from infected mice are listed in Table 1.
5. FACS tubes (5 ml polystyrene round-bottom tubes).

### 2.5 Total RNA Purification

1. TRI Reagent® (Cat. T9424, SIGMA).

### 2.6 Protein Quantification by ELISA

1. KC (CXCL1): Mouse ELISA kits.
2. MIP2 (CXCL2): Mouse ELISA kits.
3. G-CSF: Mouse ELISA kits.
4. 0.22 μM filter.

## 3 Methods

All procedures involving virulent *Y. pestis* strains should be carried out in a biological safety cabinet under biosafety level-3 (BSL-3) containment, according to the international guidelines for the use and handling of pathogenic microorganisms (http://www.cdc.gov).

### 3.1 Whole Lung Removal

1. Euthanize the *Y. pestis* intranasally infected mice by cervical dislocation or by any other methods that are approved by the Ethics Committee for Animal Experiments in your institute.
2. Attach the mouse to the sacrifice board in a dorsal position and gently spray the abdominal side with 70% EtOH to reduce fur

contamination and to keep the working environment as sterile as possible (Fig. 1a).

3. Gently remove the skin and open the rib cage to expose the lungs. Dissect the lungs by detaching them from surrounding connective tissue and carefully cut the trachea (*see* **Note 2**).
4. Wash the lungs by immersing them several times in 4 ml of cold 1× PBS. Each lung should be washed in an individual well of a 6-well plate (Fig. 1b).

### 3.2 Whole Lung Processing

In this section, lung cell-suspensions are prepared for (a) determination of neutrophil levels using FACS analysis (steps 5–19 in Subheading 3.2.1), (b) preparation of total RNA from lung cells for cDNA analysis of chemotactic inflammatory proteins (**step 5** in Subheadings 3.2.1 and 3.2.2), and (c) assessment of neutrophil chemotactic inflammatory proteins in lung supernatants by ELISA (**step** 7 in Subheadings 3.2.1 and 3.2.3).

#### 3.2.1 FACS Analysis

1. Place the washed lung on top of a 70 μm cell strainer located on a well (of a new 6-well plate) containing 2 ml 1× PBS supplemented with protease inhibitor cocktail (Fig. 1c).
2. Chop the lung on top of the filter into small pieces carefully, without damaging the filter itself.
3. Use the plunger of a 3 ml syringe to mash the chopped pieces of the lung against the mesh until the tissue disperses completely into the cold PBS (Fig. 1c). Circular and fast movements are recommended while taking care not to tear off the mesh. A residual thin transparent connective tissue remains on the mesh and a gray cloudy cell suspension is in the well, indicating a successful extraction (Fig. 1d).
4. Dilute a 10 μl sample from the cell suspension in a 1:20 ratio with trypan blue solution (10 μl + 190 μl) and count the cells under a microscope. Calculate the absolute number of cells extracted from the whole lung. The expected cell number range between 2 and 5 × $10^7$ cells per lung of a 20–25 g mouse (*see* **Note 3**).
5. *For each individual mouse*, divide the lung cell suspension into two tubes marked I (0.6 ml) and II (1.2 ml): Tube I, for FACS analysis (Fig. 1f) and tube II for RNA preparation (Fig. 1g, as depicted in Subheading 3.2.2) (*see* **Note 4**).
6. Centrifuge all the 1.5 ml tubes containing cell suspensions at 300 × *g* for 10 min at 4 °C to pellet the lung cells.
7. Combine the whole lung supernatants *of each individual mouse* from both I and II tubes by transferring the supernatants after centrifugation into a new 1.5 ml tube (marked: "mouse number" and Lung Sup., *see* **Note 5**). The combined lung

supernatants from each mouse will be used for further assessment of neutrophil chemotactic inflammatory signals secretion using ELISA (Fig. 1h) (*see* Subheading 3.2.3).

8. For RNA preparation from lung cell (tube II, **steps 5** and **6**), proceed to Subheading 3.2.2.
9. For FACS analysis of lung neutrophils re-suspend the cell pellet in tube I (**steps 5** and **6**), with 1 ml of 4% PFA and incubate at room temperature for 45–60 min (*see* **Note 6**).
10. Shake the tube gently every 15 min.
11. Centrifuge the cell-suspension of tube I in a 4 °C pre-cooled centrifuge for 10 min at 300 × *g*.
12. Gently discard the PFA solution and re-suspend the pellet of tube I in 1 ml of sterile PBS to wash away residual PFA content.
13. Re-centrifuge tube I as indicated in **step 10**.
14. Gently discard the PBS solution and re-suspend the pellet of tube I in 300–500 μl of sterile FACS buffer according to the size of the pellet. In compliance with local institutional regulations, sterile samples may be analyzed outside the BSL-3 laboratory.
15. Transfer 1–5 × $10^6$ cells from the lung cells suspension in tube I (**step 14**), to a new FACS-tube for staining and add FACS buffer to a total volume of 100 μl.
16. Add 1 μl of each one of the conjugated antibodies (APC anti-mouse Ly-6G and PerCP-Cy5.5 anti-mouse CD11b) to the new FACS-tube (containing 1–5 × $10^6$ lung cells, **step 15**), and incubate at 4 °C for 30 min. For analysis of more than one individual mouse-sample at a time, it may be beneficial to pre-mix the antibodies into one staining solution (Ab mixture) (*see* **Note 7**).
17. Wash the cells with 1 ml of FACS buffer and centrifuge the cells in a 4 °C pre-cooled centrifuge for 10 min at 300 × *g*.
18. Carefully discard the FACS buffer and re-suspend the pellet in 300–500 μl of FACS buffer.
19. Samples are now ready to be acquired by a FACS machine (*see* **Note 8**). Representative FACS analysis is shown in Fig. 2.

#### 3.2.2 Total RNA Extraction from Lung Cells

1. To pellet lung cells, centrifuge tube II containing ~1.2 ml cell-suspension (described in **step 5** Subheading 3.2.1), at 300 × *g* for 10 min at 4 °C.
2. For total RNA extraction from the lung, we recommend the use of Tri-reagent® (Sigma-Aldrich) according to the manufacturer's instructions. This purification method usually provides high RNA yields that qualify for performing quantitative reverse transcription PCR (RT-qPCR) analysis (*see* **Note 8**).

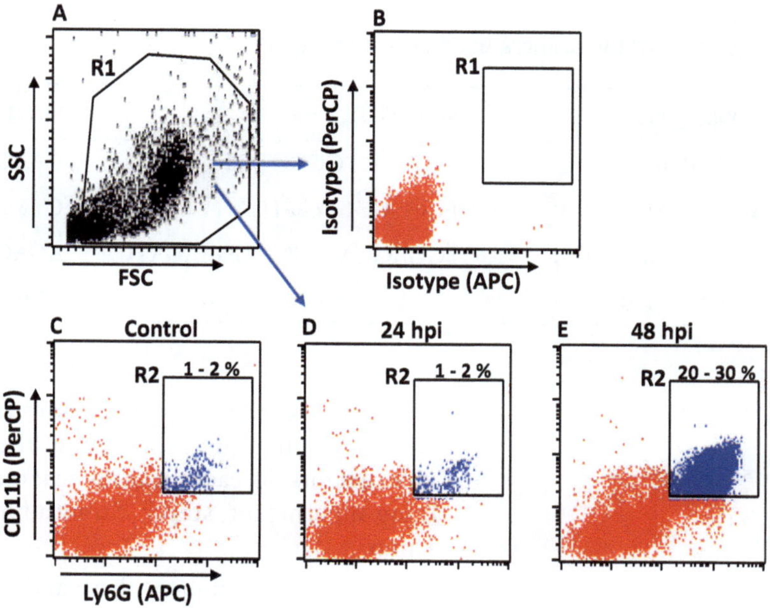


**Fig. 2** Gating strategy of flow cytometric analysis for whole lung derived neutrophils during pneumonic plague: (**a**) First gate on total cells extracted from the lung fixed with 4% PFA as described in Subheading 3.2.1. Forward scatter control (FSC) indicating cell size and side scatter control (SSC) indicating cell granularity. R1 gate refers to the relevant cells according to their size and granularity excluding any debris and doublets in the sample acquired. (**b**) Whole lung cells are stained with isotype control antibodies to position the R2 gate, referring to Ly6G- and CD11b-positive cells only. Whole lung cells extracted from control mice (**c**), mice infected for 24 h (**d**) or 48 h (**e**) with the highly virulent *Y. pestis* strain Kimberley53 (Kim53) are labeled with the neutrophil markers Ly6G (APC) and CD11b (PerCP) and are distinguished by R2 gate. Representative neutrophil percentages are indicated according to analysis results using CellQuest™ Pro software (BD Bioscience, USA)

3. Take 2 μg of total RNA for cDNA analysis of chemotactic pro-inflammatory proteins (*see* **Note 9**).
4. Perform quantitative PCR analysis using SYBR green chemistry (*see* **Note 10**). PCR conditions are as follow: Initial denaturation step 95 °C, 10 min followed by 40 cycles of PCR amplification (95 °C, 15 s and 60 °C, 60 s). Dissociation step was performed at the end of the amplification (95 °C, 15 s). The fold-change in the quantity of gene transcripts were measured and compared relative to the gene expression of the *hprt* gene using the comparative ($-2^{\Delta\Delta Ct}$) method. Primer sequences used are listed in Table 2.

#### 3.2.3 ELISA

1. Lung supernatants may be used for analysis of proteins involved in the host immune response to the infection using ELISA (e.g. [13]).

**Table 2**
**Sequences of the primers used in this study**

| Mouse gene | Forward 5′-3′ | Reverse 5′-3′ |
|---|---|---|
| KC NM_008176 | CAATGAGCTGCGCTGTCAGT | CAAGGGAGCTTCAGGGTCAA |
| MIP-2 NM_009140 | CCTGCCGGCTCCTCAGT | CTTTTTGACCGCCCTTGAGA |
| G-CSF NM_009971 | CCTGGAGCAAGTGAGGAAGATC | AGAGAGTGGCCCAGCAACAC |
| HPRT NM_013556 | AGTACAGCCCCAAAATGG | TCCTTTTCACCAGCAAGCT |

2. For analysis of the host pro-inflammatory response in the lung, centrifuge the tube containing the combined whole lung supernatants of each mouse (marked: "mouse number" and Lung supernatant, **step 7**, Subheading 3.2.1) at 15,000 rcf for 1 min.
3. Gently collect the upper phase and filter the clarified supernatant through a 0.22 μM filter to eliminate any bacterial contamination that may exist in the lungs or either rough tissue debris or precipitants that might interfere with the clarity of your sample.
4. In compliance with local institutional regulations, sterile samples may be analyzed outside the BSL-3 laboratory (*see* **Note 8**).
5. For KC (CXCL1), MIP2 (CXCL2), and G-CSF expression-level quantification, we use Mouse ELISA kits according to manufacturer's instructions.

## 4 Notes

1. The current protocol is described for C57BL/6 mice, yet it may be implemented for experimentation with any other murine strain. In our laboratory, the inbred C57BL/6 and the outbred OF1/CD1 female mice are routinely used for *Y. pestis* pathogenesis analysis.
2. Collection of blood or other tissue samples from the infected animals is optional. To do so, mice should be first anesthetized, and only when no reflexes are observed, collect the blood sample by heart puncture and proceed to harvest all other tissues as described in Subheading 3.2.1.
3. The absolute number of neutrophils in the lung is calculated by multiplying neutrophil percentage and the total number of white blood cells (WBC) obtained in **step 4** in Subheading 3.2.1.

4. The total number of *Y. pestis* bacterial loads in the lung can also be measured at this point by plating a sample of the whole lung extract on rich agar plates and incubating for 48 h at 28 °C.
5. The supernatants from lung suspensions of each individual mouse that were separated into two tubes and centrifuged to pellet the cells (I: lung cells for FACS analysis and II: lung cells for RNA analysis), are combined again in this step for further analysis by ELISA (Subheading 3.2.3).
6. Inactivation of samples using 4% PFA or filtration needs to be performed using a validated inactivation method and inactivation verified prior to removal from the BSL-3.
7. For staining of multiple cell samples collected from several individual mice, it is convenient to prepare a pre-mixed staining Ab mixture in a separate tube from which equivalent amounts are added to each FACS sample. In addition, using isotype control matched antibodies is highly recommended to subtract non-specific binding and background staining of the antibodies.
8. Inactivation of the samples from *Y. pestis* infected tissues need to be performed using a validated inactivation method and inactivation verified prior to removal from the BSL-3. During this time period, total RNA and whole lung supernatants should be stored at −70 °C and cells for FACS analysis should be stored at 4 °C. Upon approval from the local institutional regulator, sterile samples may be analyzed outside the BSL-3 facility.
9. We recommend the use of the reverse-transcribed qScript cDNA synthesis kit, according to manufacturer's instructions (Quanta Biosciences, USA).
10. We perform the analysis using an ABI 7500 instrument (Applied Biosystems, USA) with PerfeCTa SYBR Green FastMix Low ROX, according to manufacturer's instructions (Quanta Biosciences, USA).

## References


1. Craig A, Mai J, Cai S, Jeyaseelan S (2009) Neutrophil recruitment to the lungs during bacterial pneumonia. Infect Immun 77 (2):568–575
2. Strieter RM, Kunkel SL (1994) Acute lung injury: the role of cytokines in the elicitation of neutrophils. J Invest Med 42(4):640–651
3. Williams MR, Azcutia V, Newton G, Alcaide P, Luscinskas FW (2011) Emerging mechanisms of neutrophil recruitment across endothelium. Trends Immunol 32(10):461–469. https://doi.org/10.1016/j.it.2011.06.009
4. Mocsai A (2013) Diverse novel functions of neutrophils in immunity, inflammation, and beyond. J Exp Med 210(7):1283–1299. https://doi.org/10.1084/jem.20122220
5. Perry RD, Fetherston JD (1997) *Yersinia pestis*--etiologic agent of plague. Clin Microbiol Rev 10(1):35–66
6. Inglesby TV, Dennis DT, Henderson DA, Bartlett JG, Ascher MS, Eitzen E et al (2000) Plague as a biological weapon: medical and public health management. Working Group

on Civilian Biodefense. JAMA 283 (17):2281–2290
7. Kool JL (2005) Risk of person-to-person transmission of pneumonic plague. Clin Infect Dis 40(8):1166–1172
8. Pechous RD, Sivaraman V, Stasulli NM (2016) GoldmanWE. Pneumonic plague: the darker side of *Yersinia pestis*. Trends Microbiol 24 (3):190–197
9. Agar SL, Sha J, Foltz SM, Erova TE, Walberg KG, Parham TE et al (2008) Characterization of a mouse model of plague after aerosolization of *Yersinia pestis* CO92. Microbiology 154 (Pt 7):1939–1948
10. Bubeck SS, Cantwell AM, Dube PH (2007) Delayed inflammatory response to primary pneumonic plague occurs in both outbred and inbred mice. Infect Immun 75(2):697–705
11. Lathem WW, Crosby SD, Miller VL, Goldman WE (2005) Progression of primary pneumonic plague: a mouse model of infection, pathology, and bacterial transcriptional activity. Proc Natl Acad Sci U S A 102(49):17786–17791
12. Price PA, Jin J, Goldman WE (2012) Pulmonary infection by *Yersinia pestis* rapidly establishes a permissive environment for microbial proliferation. Proc Natl Acad Sci U S A 109 (8):3083–3088
13. Vagima Y, Zauberman A, Levy Y, Gur D, Tidhar A, Aftalion M, Shafferman A, Mamroud E (2015) Circumventing *Y. pestis* virulence by early recruitment of neutrophils to the lungs during pneumonic plague. PLoS Pathog 11: e1004893
14. Vagima Y et al (2012) Early sensing of *Yersinia pestis* airway infection by bone marrow cells. Front Cell Infect Microbiol 2:143
15. Cornelis GR, Wolf-Watz H (1997) The *Yersinia* Yop virulon: a bacterial system for subverting eukaryotic cells. Mol Microbiol 23 (5):861–867
16. Viboud GI, Bliska JB (2005) *Yersinia* outer proteins: role in modulation of host cell signaling responses and pathogenesis. Annu Rev Microbiol 59:69–89
17. Zauberman A, Cohen S, Mamroud E, Flashner Y, Tidhar A, Ber R et al (2006) Interaction of *Yersinia pestis* with macrophages: limitations in YopJ-dependent apoptosis. Infect Immun 74(6):3239–3250
18. Spinner JL, Cundiff JA, Kobayashi SD (2008) *Yersinia pestis* type III secretion system-dependent inhibition of human polymorphonuclear leukocyte function. Infect Immun 76:3754–3760. https://doi.org/10.1128/IAI.00385-08
19. Pechous RD et al (2013) Early host cell targets of *Yersinia pestis* during primary pneumonic plague. PLoS Pathog 9:e1003679
20. Chung LK, Bliska JB (2016) *Yersinia* versus host immunity: how a pathogen evades or triggers a protective response. Curr Opin Microbiol 29:56–62
21. Stasulli NM, Eichelberger KR, Price PA, Pechous RD, Montgomery SA, Parker JS, Goldman WE (2015) Spatially distinct neutrophil responses within the inflammatory lesions of pneumonic plague. MBio 6:e01530–e15

# Part III

## Invertebrate Models of Infection (In Vivo and In Vitro)

# Chapter 11

# Co-infection Assay to Determine *Yersinia pestis* Competitive Fitness in Fleas

**Athena Lemon, Amelia Silva-Rohwer, Janelle Sagawa, and Viveka Vadyvaloo**

## Abstract

Co-infection refers to the simultaneous infection of a host by multiple pathogenic organisms. Experimental co-infection studies using a mutant and its isogenic wild type have proven to be profoundly sensitive to analysis of pathogen factor mutation-associated fitness effects in in vivo models of infectious disease. Here we discuss the use of such co-infection experiments in studying the interaction between *Yersinia pestis* and its flea vector to more sensitively determine the critical bacterial determinants for *Y. pestis* survival, adaptation, and transmission from fleas. This chapter comprises two main sections, the first detailing how to infect fleas with mutant and wild type *Y. pestis* strains, and secondly how to process infected fleas and specifically quantify distinct *Y. pestis* strain burdens per flea. The *Y. pestis* competitive fitness co-infection model in fleas is insightful in evaluating the consequence of a mutation which may not be obvious in single-strain flea infections where there is less selective pressure.

**Key words** *Yersinia pestis*, Co-infection, Competitive fitness, Transmission factors, Flea infection

## 1 Introduction

*Yersinia pestis*, the etiological agent of plague, is transmitted to a mammalian host by the bite of a flea. Fleas initially acquire *Y. pestis* from an infected host through blood feeding. Within the flea digestive tract, *Y. pestis* preferentially colonizes the foregut proventriculus and/or midgut [1–3]. A short extrinsic incubation period follows acquisition during which *Y. pestis* adapts to the flea gut physiochemical environment and multiplies to develop into a dense cohesive biofilm in the flea digestive tract. During this process the foregut proventriculus, the valvular organ that pumps the blood meal into the midgut, can become efficiently blocked by biofilm thereby impeding its normal function. It is at this point

Athena Lemon and Amelia Silva-Rohwer contributed equally to this chapter.

Viveka Vadyvaloo and Matthew B. Lawrenz (eds.), *Pathogenic Yersinia: Methods and Protocols*, Methods in Molecular Biology, vol. 2010, https://doi.org/10.1007/978-1-4939-9541-7_11, 

that the flea is primed for a biofilm-mediated blockage regurgitative transmission event at the next blood feeding. This is the classical biological mechanism of transmission of *Y. pestis* from fleas [2, 3]. Traditionally, to determine the role a particular bacterial-encoded factor plays in formation of a transmissible infection in fleas, a single infection of fleas with either mutant or wild type bacteria is employed. The dynamics of single infection, such as rate of biofilm gut blockage formation, rate of flea infection, and bacterial burden per flea [4–6] are then compared between the mutant and wild type infections. These studies are extremely useful for the identification of major transmission factors with respect to the biofilm-mediated regurgitation blockage phenotype. Yet such single strain infections of mutants in key transmission factors have rarely, if ever, simultaneously revealed any survival defects associated with the mutation [4–6]. However, the recent use of *Y. pestis* co-infection experiments in fleas that utilize a mutant competed against an isogenic wild type strain revealed that several mutants exhibiting no survival defects during single-strain infection [6–8] incurred significant fitness defects during competitive co-infection with a wild type strain [7–10]. This illuminated that compensatory effects, in some instances, may mask identification of the true impacts of a specific *Y. pestis* factor on flea gut infection. Henceforth, flea co-infection studies have been established as a powerful and sensitive means for assessing if a mutant incurs fitness costs or benefits when competing for resources in the flea digestive tract.

In general, the enhanced sensitivity of competitive co-infection models is advantageous in the understanding of subtle, temporally modulated, and host immune system influenced factors that contribute to a pathogen's ability to infect a host. In fleas, *Y. pestis* co-infection experiments may represent a realistic experimental infection model for naturally infected wild fleas at plague foci, given that once *Y. pestis* is ingested it is thought to compete against the flea's natural microbiota to gain advantage [11]. Fleas reared and maintained under controlled laboratory conditions may not exhibit the same diversity of microbiota as wild fleas, limiting detection of an atypical fitness phenotype in a single-strain infection model [11].

The general strategy of *Y. pestis* co-infection of fleas is to infect a large cohort of fleas with a 1:1 mixture of a mutant and its isogenic wild type and calculate the ratio of the co-infected strains per flea over time. The strains need to be differentiated by a selection marker. Any selection marker can be used for one of the strains as long as it is verified that this marker does not confer any fitness effect to the strain. In this chapter, we use antibiotic resistance to selectively distinguish between two co-infecting strains. The fleas are fed a blood meal with an equally proportioned mixture of co-infecting strains. The total bacterial load must contain the minimum infectious dose for colonization and maintenance of a persistent infection of the digestive tract of the *Xenopyslla cheopis* flea species, which is $5 \times 10^8$ to $1 \times 10^9$ CFU/mL [12]. Fleas are collected immediately after infected blood meal acquisition and at different time points post-infection. The contents of a single co-infected flea are duplicate plated on rich media generally selective for *Y. pestis* with and without antibiotic selection, allowing enumeration of the antibiotic resistant strain alone, and total *Y. pestis* per flea respectively. The percentage of each strain in the flea gut can subsequently be determined from fleas collected at different time points post-infection.

An example of a flea co-infection study in which a mutant in the known critical *Y. pestis* transmission factor, PhoP [6], and a wild type strain bearing an antibiotic resistance cassette are co-infected is depicted in Fig. 1. A wild type *Y. pestis* KIM6+ strain genetically marked with a kanamycin resistance cassette in the non-deleterious *glmS-pstS* site on the chromosome is used as the selective antibiotic strain (KIM6+ *glmS-pstS*::*kan*$^R$ also called KIM6+::*kan*$^R$, unpublished data) and represents the isogenic wild type strain that will be competed with the KIM6+ Δ*phoP* mutant [6]. First a confirmation that the KIM6+::*kan*$^R$ does not confer a fitness impact was made by competing the strain with the isogenic *Y. pestis* KIM6+ wild type strain. The subsequent co-infection experiment of the KIM6+::*kan*$^R$ and Δ*phoP* mutant [6] demonstrates a significant survival fitness defect of the Δ*phoP* mutant over a week, and this defect is not resolved by 4 weeks post-infection (Fig. 1). This survival fitness defect exhibited by the Δ*phoP* mutant when co-infected with the wild type recapitulates the observation of a published co-infection study using another PhoP mutant with a wild type strain harboring a plasmid encoding an ampicillin resistance cassette [13]. Generally, a complemented mutant strain should be co-infected with the same wild type strain to verify that this phenotype can be restored in a separate co-infection experiment; such data is not demonstrated here.

One potential shortcoming that experimenters may want to be mindful of when infecting with two strains into the same flea is the possibility of synergistic or *trans*-complementation effects in which no discernable difference in infectivity is observed. In such a case, the predicted or known biological function of the mutated factor might provide a clue as to whether this may be a significant consideration.

It is assumed that the researcher has the infrastructure, necessary equipment and knowledge of flea husbandry, so these aspects of the experimentation will not be covered. The experiments described here are performed with biosafety level two designated *Y. pestis* strains, devoid of the pCD1 plasmid. Appropriate biosafety and regulatory practices will have to be considered if fully virulent strains are to be used.

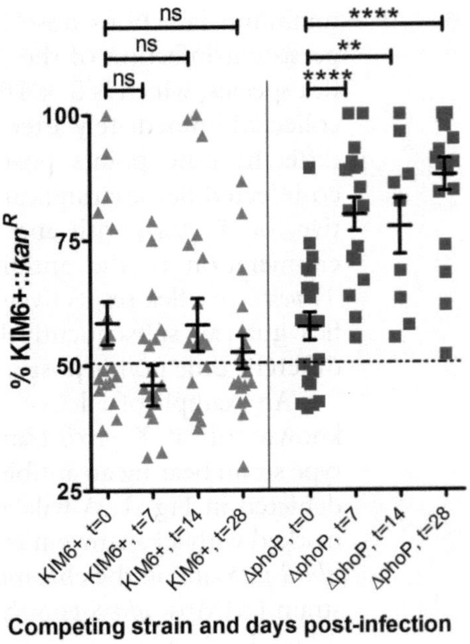


**Fig. 1** Mutation in PhoP confers a competitive fitness defect that is apparent when a PhoP mutant is co-infected with a *Y. pestis* KIM6+ wild type strain in fleas. In two separate co-infection experiments, *X. cheopis* fleas were co-infected with either a 1:1 ratio of *Y. pestis* KIM6+::*kan*$^R$ and KIM6+ (triangles) or, KIM6+::*kan*$^R$ and KIM6+ Δ*phoP* (squares). The *Y. pestis* bacterial loads were determined for each of 10–20 fleas at time points 0 (1 h following acquisition of the co-infected blood meal), 7, 14 and 28 days post-infection. The percentage of KIM6+::*kan*$^R$ in the co-infection was determined and plotted for each experiment and time point. Error bars represent mean ± SEM. A one-way ANOVA and *post-hoc* Bonferroni test were used to determine statistical significance between the time point 0 ($t = 0$) and other time points for each experiment

## 2 Materials

### 2.1 Skinning Mice

1. 6+ week old mice (*see* **Note 1**).
2. Sterile 1× Phosphate-Buffered Saline (1× PBS).
3. Corkboard with push pins.
4. Plastic spoon.
5. Rubber bands.
6. Cling film/Saran wrap.
7. 70% ethanol (EtOH).

### 2.2 Bacterial Culture

1. Congo Red (CR) agar plates [14].
2. Brain Heart Infusion (BHI) broth.
3. 250 mL sterile glass Corning bottle.

4. 250 mL sterile centrifuge tubes and centrifuge set to 4 °C.
5. *Yersinia pestis hmsHFRS+* strains (*see* **Note 2**).
6. Shaking incubator set to 26 °C and 200 rpm.
7. 28 °C incubator.
8. 37 °C incubator.

### 2.3 Flea Infection and Sorting

1. ~300 *Xenopsylla cheopis* fleas per infection (*see* **Note 3**).
2. Mouse blood (*see* **Note 4**).
3. Water-jacketed glass membrane feeder [15] (*see* **Note 5**).
4. Water bath set at 42 °C with tubing to attach to water-jacketed feeder.
5. Peristaltic pump.
6. Flea cages (*see* **Note 6**).
7. Ultra-fine tip forceps, 0.09 × 0.05 mm tip.
8. Flea vacuum apparatus.
9. Dissecting microscope.
10. 8–10 in. tall basin.
11. Chill table.
12. Waterproof medical tape.
13. Electric clippers with 0.25 mm clipper blade.
14. Glass petri dish.
15. Columbia blood agar plates.
16. Ice bucket.
17. Insect incubator set at 75% relative humidity and 21 °C.

### 2.4 Determining CFU/Flea

1. Brain Heart Infusion (BHI) agar plates.
2. Two water baths set to 42 °C and 46 °C (*see* **Note 7**).
3. BHI broth.
4. 0.7% soft molten BHI agar: Melt 100 mL BHI agar completely in a microwave oven, immediately add an equal volume of BHI broth to it and incubate in the water bath set to 46 °C.
5. 10 mg/mL hemin stock solution. Dissolve 10 mg hemin in 1 mL 0.05 N sodium hydroxide. Store at 4 °C.
6. 1 mg/mL irgasan stock solution: Dissolve 1 mg irgasan in 1 mL 100% EtOH. Store at −20 °C.
7. Selective antibiotic (e.g. 50 mg/mL kanamycin) (*see* **Note 8**).
8. Sterile 1× PBS.
9. 95% EtOH.
10. 3% Hydrogen peroxide ($H_2O_2$).

11. 100 × 15 mm sterile polystyrene petri dish.
12. Lysing Matrix H tubes (MP Biomedicals) (*see* **Note 9**).
13. Bunsen burner.
14. Two ultra-fine tipped forceps, 0.09 × 0.05 mm tip.
15. Bead beater (e.g. MP Biomedicals Fast Prep®-24).
16. Glass tubes with caps.

## 3 Methods

### 3.1 Culturing *Y. pestis* for Infection

1. Pick a 2 mm patch of single red-pigmented *Y. pestis* colonies from a CR plate (*see* **Note 2**) and culture, with antibiotic if appropriate, in 3 mL BHI broth overnight, with agitation.
2. Transfer entire volume of overnight culture into sterile 250 mL glass bottle containing 100 mL of BHI broth, with antibiotic if appropriate, and incubate overnight at 37 °C, without agitation.

### 3.2 Preparing Mouse Skin

1. Euthanize a 6+ week old mouse in accordance with IACUC approved protocols (*see* **Note 10**).
2. Spray mouse with 70% EtOH to sterilize the skin surface.
3. Use scissors to cut the skin around the neck of the mouse. Then, cut the skin from the base of the skull, down the backbone of the mouse, to the tail.
4. Gently remove skin from the mouse and retain for experiment. Discard mouse carcass appropriately.
5. Place the skin, fur-side down, on corkboard. Using push pins, tautly stretch the skin and attach by the far corners to the board (*see* **Note 11**).
6. Put a few drops of 1× PBS onto the skin surface and scrape off the fat layer of the skin using the side edge of a disposable plastic spoon (*see* **Note 12**).

### 3.3 Preparing Artificial Glass Membrane Feeder (See Fig. 2)

1. Prepare the glass feeder by tautly stretching the mouse skin over the large 62 mm flange opening, fur-side out. Secure skin onto the apparatus with rubber bands.
2. Using electric clippers, closely shave fur to approximately 1–2 mm length. Assess for leaks in the skin by adding 1× PBS into the skin covered chamber.
3. Attach feeder to the tubing of the peristaltic pump. Switch on pump to allow circulation of water to pervade outer jacket of feeder in order to maintain inner chamber at 37 °C (*see* **Note 13**).

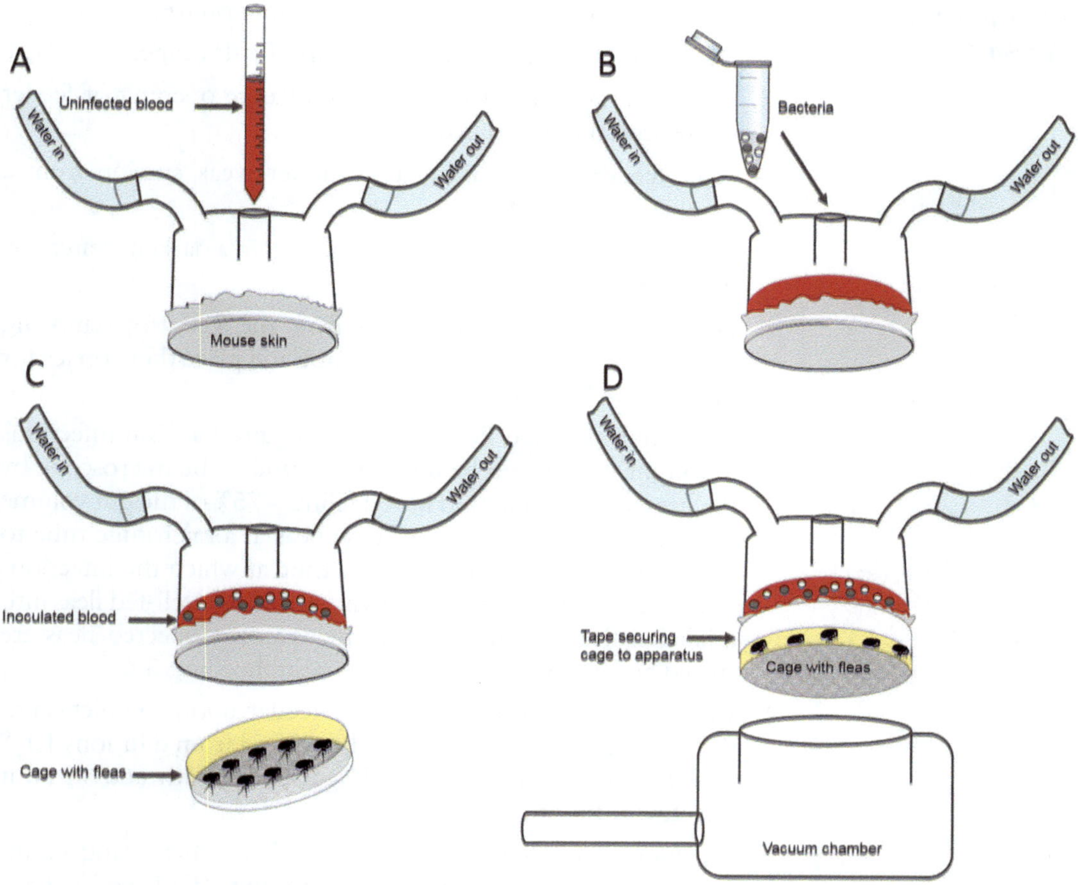


**Fig. 2** Artificial flea feeding schematic. (**a**) Uninfected blood is added to the opening of the water-jacketed feeder with mouse skin to warm. (**b**) A 1:1 ratio of mutant strain (open circles) and isogenic wild type (grey circles) are added to blood. (**c**) The cold immobilized cohort of fleas is placed in the flea cage. (**d**) The flea cage is secured to the fur-side of the skin on the feeder, then attached to the vacuum chamber for prolonged feeding

4. Spin down the bacterial culture at 7500 × *g* at 4 °C and resuspend the pellet in 500 μL of sterile 1× PBS and keep on ice.
5. Determine concentration of bacteria. Calculate how much is needed for a final concentration of $5 \times 10^8$ to $1 \times 10^9$ bacteria/mL blood separately for each strain (*see* **Note 14**).
6. Add 5 mL of blood to the inner chamber of the feeder to be pre-warmed approximately 5–10 min prior to bacterial inoculation.
7. Add equivalent amounts of each strain to 5 mL of pre-warmed blood in the feeder. Swirl blood and bacteria to properly mix together within feeder (*see* **Note 15**).

### 3.4 Infecting Fleas (See Fig. 2)

1. Chill fleas to immobilize for at least 5 min on ice.
2. Place immobile fleas into the bottom of a flea cage.
3. Securely attach cage to skin covered flange opening of feeder with surgical tape (*see* **Note 16**).
4. Place feeder with attached cage under weak suction from a vacuum (*see* **Note 17**).
5. Allow fleas to feed for approximately 1 h in a darkened area (*see* **Note 18**).
6. Disassemble and detach the cage from the feeder by removing the surgical tape, collect fleas in a tube and place them on ice for 10 min (*see* **Note 19**).
7. Select only well-fed fleas that have acquired a fresh infectious blood meal. This can be determined under the microscope by the presence of bright red blood filling ≥75% of the gut volume (*see* **Note 20**). Collect 20 of these fleas in a microfuge tube to represent the time point of day 0; time at which the infectious blood meal was acquired. Place the rest of the well-fed fleas into a labeled flea cage in an insect incubator. Collected fleas are stored at −80 °C until processed in Subheading 3.6.
8. Collect 100 μL of blood from the feeder upon completion of flea feeding. Serially dilute and plate blood from dilutions $10^{-6}$ to $10^{-7}$ onto Columbia blood agar plates with and without selective antibiotic.
9. Incubate plates for ~48 h in a 28 °C incubator and count colonies to determine colony forming units (CFU) of *Y. pestis* per mL blood. Verify CFU is within desired infectious dose range and that both strains are equally represented.
10. Fleas should be fed twice a week on a fresh blood meal for the duration of the experiment according to established lab protocols for flea maintenance. Collect fleas at required time points post-infection and store at −80 °C until processing according to Subheading 3.6 (*see* **Note 21**).

### 3.5 Preparing Top Layer Molten Agar

The following materials are enough for processing 10 co-infected fleas.

1. Allow 20 BHI plates to warm to room temperature (*see* **Note 22**).
2. Place two separate bottles of 100 mL melted 0.7% BHI soft agar in a 46 °C water bath (*see* **Note 23**).
3. Label 10 Lysing Matrix H tubes, 1 through 10, and sterilely add 1 mL of 1× PBS to each tube.
4. Place 20 capped glass test tubes into a 42 °C water bath.

5. Add hemin and irgasan to a final concentration of 10 μg/mL and 1 μg/mL, respectively, to each of the two molten agar bottles and swirl to mix.
6. Add appropriate selective antibiotic to one of the bottles and label.
7. Sterilely add 4 mL of the molten soft agar without selective antibiotic to 10 warmed glass test tubes resting in the water bath. Repeat with another 10 warmed glass test tubes using the molten agar containing selective antibiotic. Keep tubes in the 42 °C water bath.

### 3.6 Sterilizing Flea Surface and Preparing Flea Triturate

Fleas must be surface sterilized prior to processing to eliminate surface contaminants, and specifically quantify *Y. pestis* in the flea digestive tract. The following methods will be carried out using aseptic technique. Light a Bunsen burner to create a sterile environment on the laboratory bench.

1. Retrieve 10 co-infected fleas from −80 °C freezer storage.
2. Add 1 mL of 3% $H_2O_2$ to tube containing fleas and incubate for 2 min. Mix gently by inverting tube 2-3 times during incubation.
3. Gently draw off the $H_2O_2$ with a pipetman.
4. Add 1 mL of 95% EtOH to tube containing fleas and incubate for 1 min. Mix gently by inverting tube 2–3 times during incubation.
5. Gently draw off EtOH using a pipetman and add another 1 mL of 95% EtOH to tube containing fleas.
6. Pour fleas and EtOH onto a sterile petri plate. Draw off as much EtOH by tipping plate to a 45° angle to collect EtOH.
7. Separate the fleas using flame sterilized fine tipped forceps. Allow fleas to air dry in the petri dish (*see* **Note 24**).
8. Rinse fleas once by adding 10 mL sterile distilled water to petri dish. Gently remove water with a pipet.
9. Separate the fleas with sterilized forceps as before and allow them to air dry again for ~10 min.
10. Carefully transfer one flea into each of the 10 labeled Lysing Matrix H tube using sterilized forceps.
11. Lyse and homogenize fleas using a Fastprep bead beater (*see* **Note 25**). Machine settings: one cycle of 20 s at speed 6 (*see* **Note 26**).

### 3.7 Plating Flea Triturate (See Fig. 3)

1. Mix contents of crushed fleas by inverting tube 3–4 times.
2. Transfer 25 μL of crushed flea triturate to one glass tube containing molten agar without selective antibiotic (*see* **Note 27**).

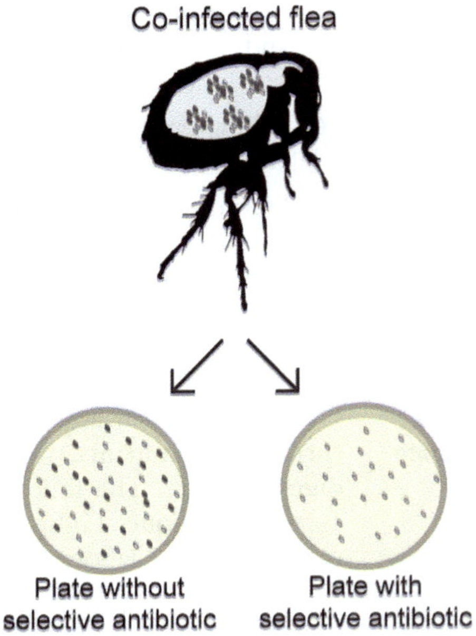


**Fig. 3** Co-infected flea plating schematic. The flea triturate is mixed with molten agar and plated as a top layer on an agar plate. For molten agar containing no selective antibiotic, both strains (open circles for antibiotic resistance marker encoding strain, closed circles for unmarked strain) grow providing a total CFU count per flea while selective antibiotic containing media allows for CFU enumeration of the strain that contains an antibiotic selective marker

3. Vortex the molten agar containing flea triturate, briefly flame the mouth of tube and pour entire contents onto the center of one appropriately labeled BHI agar plate. Rock the plate gently to distribute the agar evenly over the surface.
4. Set the plate on a flat surface and repeat for the remaining nine samples using molten agar tubes without selective antibiotic. Allow the plates to solidify for approximately 15 min.
5. Repeat **steps 2–4** for selective antibiotic medium (*see* **Note 28**).
6. Add 3–4 mL of remaining molten soft agar (with or without selective antibiotic as appropriate) to overlay on each plate.
7. Allow the overlay layer to solidify for approximately 15 min.
8. Invert and incubate plates at 28 °C for 48–72 h.
9. Count colonies on plates using colony counter and multiply by the dilution factor to determine CFU *Y. pestis* per flea (*see* **Note 29**).
10. For each flea, calculate the percent selective marker strain over the total CFU per flea by dividing the selective antibiotic CFU count over total CFU count and multiply by 100. Plot and compare these percentages between different days post-

infection. Apply statistical analyses. This should identify a significant advantage or disadvantage in fitness between compared strains (Fig. 1).

## 4 Notes

1. It is easier to see fleas on white-furred mice, i.e. Swiss Webster or CD-1.
2. For flea infections, *hmsHFRS+ Y. pestis* strains are usually used because these strains form a biofilm, which is important for persistent colonization of the flea gut in *X. cheopis* fleas [4]. Picking colonies from CR agar plates allows for confirmation of this phenotype prior to flea infection.
3. Co-infection studies described here are optimized for infection of *X. cheopis* fleas for which the dynamics of persistent *Y. pestis* infection has been well-established. Not all fleas may exhibit persistent colonization of the digestive tract by *Y. pestis*.
4. The blood is treated with sodium-heparin, but defibrinated blood can also be used [1]. Commercially purchased blood should be used as fresh as possible, i.e. within ~24–48 h of arrival.
5. The glass feeder is a patented invention (https://patents.google.com/patent/US4850305). The outer chamber forms the water jacket through which water at 42 °C can be circulated. This heated water can maintain the blood that is filled into the inner chamber at the physiologically relevant temperature of 37 °C. This feeder is commercially available through Lillie Glass Blowers, Inc. (Smyrna, USA).
6. Flea cages are custom made with a 62 mm diameter opening on the upper side that matches the diameter of the flange on the glass membrane feeder. A fine-pored nylon breathable mesh composes the base of the cage and top of the lid of the cage. Infected fleas can be maintained in the cage or a pint Mason glass jar lined at the bottom with a thin layer of sawdust. Instead of the solid aluminum flat cap on the Mason jar, a fine-pored mesh is glued to the screw band to prevent flea egress while maintaining airflow.
7. The water bath temperature must provide a balance of preventing molten agar from solidifying and death of *Y. pestis*.
8. We have successfully used 100 μg/mL carbenicillin for selection with strains harboring an ampicillin resistance cassette. However, each antibiotic used should be pre-tested to ensure absence of synergistic activity with irgasan.

9. Lysing Matrix H tubes have 2 mm glass and 2 mm zirconium beads that break through the flea's chitinous exoskeleton, disrupt the midgut tissue, and break up the biofilm matrix without causing bacterial death under the optimized settings used.
10. The IACUC approved euthanization method used should keep the skin intact.
11. Make sure not to pierce or tear the main surface area of the skin.
12. The spoon should glide over the skin, there is no need for excessive dousing with PBS. Use enough force to scrape off the fat, without tearing the skin. Skins are typically prepared a day ahead and store at −20 °C wrapped in Saran wrap inside a sealed plastic container.
13. The 42 °C temperature of the water is sufficient to ensure that blood temperature is maintained at 37 °C. Blood is warmed within 10 min and ready for inoculation with bacteria.
14. This can be calculated using a Petroff-Hausser counting chamber. A pre-determined $OD_{600}$ reading that represents the required *Y. pestis* CFU can also be used.
15. Blood may also be pre-warmed in a 37 °C incubator for 10 min before adding bacteria. Once bacteria are added, the inoculum should be mixed well and then placed into the feeder.
16. We use surgical tape to attach the bottom of the flea cage to the skin covering the feeder. The surgical tape adheres better to the mouse skin, and prevents tearing during disassembly of the feeder after infection.
17. The light suction from the vacuum pulls the blood against the mouse skin, and creates a humid micro-environment ideal for fleas feeding that mimics natural feeding conditions on a live host.
18. We cover the entire vacuum system and feeder with attached fleas with an opaque cover, like a black garbage bag.
19. When fleas are first placed on ice, simultaneously place a glass petri dish on the chilling table under the microscope to pre-chill in preparation to sort immobile fleas.
20. Fleas that have not fed well will sometimes clear the infection because they have not acquired an appropriate infectious dose. Such fleas will not be useful for your study and should be discarded.
21. We feed fleas on days 5, 8, 12, 15, 19, 22, and 26 post-infection. We collect fleas to analyze on day of infection and days 7, 14, and 28 post-infection.
22. Label plates 1 through 10 that normal agar will be added to and another 10 plates labeled 1 through 10 for adding agar

containing selective marker antibiotic. The numbers of these plates should correspond to the same flea triturate tube.

23. If using solidified BHI soft agar, melt in the microwave by placing bottle of agar in a pan of water and microwave to prevent boiling over of agar.
24. Be careful not to pierce fleas because this may allow infiltration of EtOH, resulting in bacterial death.
25. If using another type of lysing tube or bead beater, optimization may be required to assure adequate crushing of flea without lysing bacteria.
26. Check each tube to make sure the flea has been crushed open to expel gut contents.
27. For fleas collected on day 0, transferring 25 μL flea triturate directly from the lysis matrix tube constitutes a 1:40 dilution of the flea triturate. The same is done for fleas collected at ≥7 days post-infection, except that flea triturate is diluted 1:10 in 1× PBS first to obtain a final 1:400 dilution. This dilution series usually captures the countable range of colonies.
28. It is important to match a specific flea to a single number because the CFU calculated from the plate without antibiotics will be subtracted from the CFU from the corresponding plate with antibiotics.
29. Plates will frequently have many colonies that can be more accurately counted using a colony counter, i.e. Reichert™ Darkfield Quebec™ Colony Counter.

## Acknowledgments

This work was supported by the National Institute of Allergy and Infectious Diseases of the National Institutes of Health Grants 1R21AI097974-01 and 1R01 AI117016-01A1 to V.V.

### References

1. Bland DM, Jarrett CO, Bosio CF, Hinnebusch BJ (2018) Infectious blood source alters early foregut infection and regurgitative transmission of *Yersinia pestis* by rodent fleas. PLoS Pathog 14(1):e1006859. https://doi.org/10.1371/journal.ppat.1006859
2. Bacot AW, Martin CJ (1914) LXVII. Observations on the mechanism of the transmission of plague by fleas. J Hyg 13(Suppl):423–439
3. Jarrett CO, Deak E, Isherwood KE, Oyston PC, Fischer ER, Whitney AR, Kobayashi SD, DeLeo FR, Hinnebusch BJ (2004) Transmission of *Yersinia pestis* from an infectious biofilm in the flea vector. J Infect Dis 190(4):783–792. https://doi.org/10.1086/422695
4. Hinnebusch BJ, Perry RD, Schwan TG (1996) Role of the *Yersinia pestis* hemin storage *(hms)* locus in the transmission of plague by fleas. Science 273(5273):367–370
5. Sun YC, Koumoutsi A, Jarrett C, Lawrence K, Gherardini FC, Darby C, Hinnebusch BJ (2011) Differential control of *Yersinia pestis* biofilm formation *in vitro* and in the flea vector by two c-di-GMP diguanylate cyclases. PLoS One 6(4):e19267. https://doi.org/10.1371/journal.pone.0019267

6. Rebeil R, Jarrett CO, Driver JD, Ernst RK, Oyston PC, Hinnebusch BJ (2013) Induction of the *Yersinia pestis* PhoP-PhoQ regulatory system in the flea and its role in producing a transmissible infection. J Bacteriol 195 (9):1920–1930. https://doi.org/10.1128/JB.02000-12
7. Vadyvaloo V, Hinz AK (2015) A LysR-type transcriptional regulator, RovM, senses nutritional cues suggesting that it is involved in metabolic adaptation of *Yersinia pestis* to the flea gut. PLoS One 10(9):e0137508. https://doi.org/10.1371/journal.pone.0137508
8. Rempe KA, Hinz AK, Vadyvaloo V (2012) Hfq regulates biofilm gut blockage that facilitates flea-borne transmission of *Yersinia pestis*. J Bacteriol 194(8):2036–2040. https://doi.org/10.1128/JB.06568-11
9. Fukuto HS, Svetlanov A, Palmer LE, Karzai AW, Bliska JB (2010) Global gene expression profiling of *Yersinia pestis* replicating inside macrophages reveals the roles of a putative stress-induced operon in regulating type III secretion and intracellular cell division. Infect Immun 78(9):3700–3715. https://doi.org/10.1128/IAI.00062-10
10. Aoyagi KL, Brooks BD, Bearden SW, Montenieri JA, Gage KL, Fisher MA (2015) LPS modification promotes maintenance of *Yersinia pestis* in fleas. Microbiology 161 (Pt 3):628–638. https://doi.org/10.1099/mic.0.000018
11. Jones RT, Vetter SM, Montenieiri J, Holmes J, Bernhardt SA, Gage KL (2013) *Yersinia pestis* infection and laboratory conditions alter flea-associated bacterial communities. ISME J 7 (1):224–228. https://doi.org/10.1038/ismej.2012.95
12. Lorange EA, Race BL, Sebbane F, Hinnebusch BJ (2005) Poor vector competence of fleas and the evolution of hypervirulence in *Yersinia pestis*. J Infect Dis 191(11):1907–1912. https://doi.org/10.1086/429931
13. Fukuto HS, Vadyvaloo V, McPhee JB, Poinar HN, Holmes EC, Bliska JB (2018) A single amino acid change in the response regulator PhoP acquired during *Yersinia pestis* evolution affects PhoP target gene transcription and polymyxin B susceptibility. J Bacteriol 200(9): e00050-18. https://doi.org/10.1128/JB.00050-18
14. Surgalla MJ, Beesley ED (1969) Congo red-agar plating medium for detecting pigmentation in *Pasteurella pestis*. Appl Microbiol 18 (5):834–837
15. Wade SE, Georgi JR (1988) Survival and reproduction of artificially fed cat fleas, *Ctenocephalides felis* bouche (Siphonaptera: Pulicidae). J Med Entomol 25:186–190

# Chapter 12

## *Drosophila* as a Model for Understanding the Insect Host of *Yersinia pestis*

**Anna B. Ludlow, Cassandra D. Pauling, Melanie M. Marketon, and Deborah M. Anderson**

### Abstract

With the limited availability of genomic sequence information and no established methods for genetic knockdowns or the creation of transgenic fleas and flea cell lines, we have adopted *Drosophila melanogaster* as a model for the study of the insect life cycle of *Yersinia pestis*. Infection of *Drosophila* larvae can be used to model early colonization of fleas, while the established embryonic cell lines can be used to model insect--pathogen interactions that underlie the unique capacity of *Y. pestis* to colonize the gut of its flea host. In this chapter, we present the methods we developed for infection of *Drosophila* in vivo and in vitro.

**Key words** *Drosophila*, S2 cells, Flea, *Yersinia pestis*, Transmission, Bubonic plague

## 1 Introduction

Fleas are inefficient vectors, but transmit a handful of diseases to humans and animals including plague [1]. There are over 1000 well-documented species of fleas and another 600 that have only been recorded a single time [2]. Fleas are diverse, differing in anatomical features and vertebrate hosts as well as the ecological conditions under which they thrive. There are currently no publicly available full genome sequences of any flea species. This makes laboratory fleas generally unamenable to transcriptome analysis or site-directed mutagenesis, and thus our understanding of the insect host of *Y. pestis*, or any other flea-borne pathogen, the innate immune response to infection, and the important interactions between insect and pathogen that facilitate transmission is very limited.

For nearly 100 years, mice have been used as a surrogate for understanding human biology. Even in today's post-human

Anna B. Ludlow and Cassandra D. Pauling contributed equally to this work.

Viveka Vadyvaloo and Matthew B. Lawrenz (eds.), *Pathogenic Yersinia: Methods and Protocols*, Methods in Molecular Biology, vol. 2010, https://doi.org/10.1007/978-1-4939-9541-7_12, © Springer Science+Business Media, LLC, part of Springer Nature 2019

**Table 1**
**Genomic comparisons between *Drosophila melanogaster* and *Xenopsylla cheopis*[a]**

| Gene/gene family | % Identity |
|---|---|
| Ribosomal protein L13 | 63% |
| Hexokinases | 38–53% |
| Attacins | 20–45% |
| Defensins | 43–52% |
| Dual oxidase maturation factors | 47–49% |

[a]*X. cheopis* sequence obtained from partial transcriptome compared to *D. melanogaster* sequence using BLAST (NLM/NCBI) [13, 14]

genome research, mice continue to be essential for mechanistic studies that enable discoveries and the development of new medicines. At the nucleotide level, approximately 40% of the human genome can be aligned to the mouse genome [3]. Fleas are related to flies as mice are to humans, leading us to investigate whether *Drosophila melanogaster* could serve as a model for understanding mechanisms of insect–pathogen interactions that facilitate flea transmission of disease. Sequence comparison of select genes from *Drosophila melanogaster* and *Xenopsylla cheopis* shows 20–63% homology (Table 1).

Mechanisms of innate immunity differ in some respects and it is generally accepted that the innate immune system of fleas is less complex than flies. However, there is strong conservation of basic mechanisms for recognizing and responding to bacterial infection, such as the Toll receptors and components of the signaling pathways that lead to the production of effector molecules such as antimicrobial peptides (AMPs) and reactive oxygen species (ROS) [4].

To demonstrate proof of concept, we developed an in vivo model wherein *Y. pestis* is fed to *Drosophila* larvae and subsequently colonizes the foregut during the 4-day span of the larval phase [5]. Following infection, we showed that *Y. pestis* grows logarithmically in the larval foregut, similar to its growth within fleas. Furthermore, without the *hms* genes, *Y. pestis* was cleared from *Drosophila*, as it is in fleas. These data suggest that the requirements for biofilm are shared between *Drosophila* and fleas. In addition, genetic data support the importance of AMP and ROS resistance in the larval foregut. Therefore, like mice and men, *Drosophila* larvae and fleas have similar responses to *Y. pestis* infection. By using the powerful genetics, available reagents, and ease of laboratory rearing of *Drosophila*, we can make rapid progress in understanding the interactions between *Y. pestis* and its insect host in order to begin to understand how fleas become vectors of infectious diseases.

In addition to in vivo studies, we have developed methods for infecting *Drosophila* cells in vitro, as there are no immortalized cell lines of fleas or methods for culturing of primary cells. Previous work showed that *Y. enterocolitica* infection of Schneider 2 (S2) cells results in intracellular growth of *Y. enterocolitica*, a phenotype that required the chromosomal type III secretion system [6]. From these data, we reasoned that infection of S2 cells by *Y. pestis* may reveal a potential role for intracellular growth in the insect life cycle. For these reasons, we developed methods for in vitro infection of S2 cells and their use as a model for studying the responses of insect cells that may lead to a permissible infection and ultimately transmission from fleas. S2 cells are immortalized, commercially available *Drosophila* cells that were derived from embryonic epithelial cells [7]. These cells have properties that resemble mammalian macrophages, including their activation by bacterial infection [8]. S2 cells have been shown to be phagocytic and to produce AMPs and ROS, making them ideal for in vitro studies to understand how *Y. pestis* stimulates and modulates insect innate immune signaling [9–11]. They also grow well in culture, making this system very amenable to laboratory work. Below we present methods for reproducibly infecting *Drosophila* S2 cells in vitro as well as larvae in vivo with *Y. pestis*.

## 2 Materials

### 2.1 Bacterial Culture

1. Heart infusion agar (HIA) and heart infusion broth (HIB).
2. Congo Red Plates (HIA supplemented with 0.005% Congo Red, 0.2% Galactose).
3. *Yersinia pestis* strains (pCD1-) (*see* **Note 1**).
4. Spectrophotometer.
5. Plastic cuvettes for spectrophotometer measurements.
6. *Yersinia* selective agar (YSA).

### 2.2 Molasses Egg Collection Plates (Recipe for 20 Plates, 60 mm Diameter)

1. Ingredients:

| | |
|---|---|
| dd$H_2O$ | 270 mL |
| Agar | 1.6 g |
| Tegosept | 0.38 g |
| Baker's Yeast | 3.3 g |
| Cornmeal | 16.3 g |
| Molasses | 20 mL |

2. Mix agar and Tegosept in 200 mL $H_2O$. Boil until agar is dissolved.
3. Dissolve yeast in 30 mL $H_2O$, then stir in cornmeal.
4. Add molasses to agar, then the yeast/cornmeal/water mix.
5. Add remaining 40 mL $H_2O$. Boil for 5–10 min, then cool to 50–55 °C and dispense into 60 mm round Petri dishes.

### 2.3 Larva Collection Materials

1. *Drosophila* media (*see* **Note 2**).
2. Fly rearing chambers (vials and flasks) with plugs.
3. Embryo collection cage (*see* **Note 3**).
4. 34% Baker's yeast (2×) in water.
5. 70 μm cell strainers.
6. Small, soft-bristled paintbrushes.
7. Undiluted bleach in squirt bottle.
8. Water in squirt bottle.
9. 1% bacto-agar in $H_2O$, autoclaved.
10. Petri dishes (60 mm and 100 mm diameter round).
11. Microtiter dishes with lid (6- or 12-wells), filled with 1% agar.
12. Sealing membranes, such as Breath Easier membranes, for multi-well plates.
13. Larva incubation chamber (box with lid, large enough to house microtiter dish and wet paper towels).
14. Sterile PBS.
15. 100% Ethanol.
16. Microcentrifuge tubes.
17. Disposable microcentrifuge pestles.

### 2.4 Cell Culture

1. Schneider 2 cells (American Type Culture Collection, Manasses, VA).
2. Schneider's *Drosophila* medium with L-glutamine.
3. Fetal bovine serum (FBS).
4. Pen/strep (10 U/mL penicillin, 10 μg/mL streptomycin) (*see* **Note 4**).
5. T-25 and T-75 vented tissue culture flasks.
6. Dimethyl sulfoxide (DMSO).
7. 30% glucose (sterile).

8. Cell scrapers.
9. Trypan blue.
10. Hemacytometer.

## 3 Methods

### 3.1 Fly Rearing and Maintenance

1. Obtain wild type and mutant *Drosophila* stocks from available resources (*see* **Note 5**).
2. Passage flies to fresh media in vials every 2–3 weeks, depending on fly density and strain (*see* **Note 6**).
3. Tap bottle or vial on benchtop to knock adult flies to bottom of container.
4. Quickly remove cap and replace with inverted new vial or bottle.
5. To transfer adult flies, invert and tap the new vial or bottle on benchtop to knock flies into it. Quickly recap to prevent escape.
6. To prepare for infection, expand to bottles.

### 3.2 *Yersinia* Infection of *Drosophila* Larvae

1. Streak out *Y. pestis* strains onto HIA-Congo Red [12]. Incubate 2–3 days at 26 °C.
2. Place flies in embryo collection cage with molasses plate to acclimate overnight at 22–26 °C (Fig. 1a, b).
3. To initiate egg laying, make a thick paste of heat-killed yeast (*see* **Note 7**) and put about 500 μL of it in the middle of a fresh molasses plate (Fig. 1c).
4. Place plate in collection chamber (replacing the previous), incubate overnight at 22–26 °C.
5. On the morning of the infection, pour 1% agar into each well of a microtiter dish (6- or 12-well).

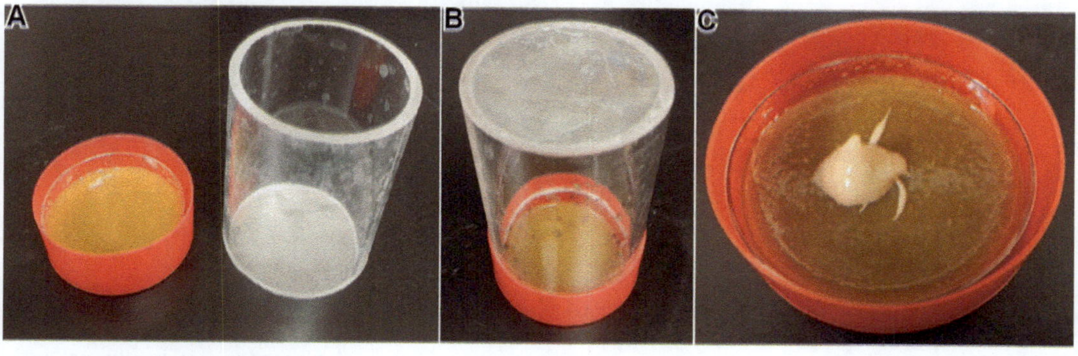


**Fig. 1** Typical embryo collection chamber for *Drosophila* research. (**a**) Empty chamber with molasses plate placed; (**b**) assembled chamber with flies; (**c**) Molasses plate with yeast paste

6. To harvest eggs (*see* **Note 8**), disassemble the embryo collection chamber. Small, but visible eggs should be covering the molasses plate. Take care when removing plate from chamber.
7. Add a small layer of water (~2–3 mL) to the plate.
8. Gently loosen eggs from the surface with a soft bristled paintbrush.
9. Tilt the plate over a cell strainer and gently squirt water over the surface of the plate to flush eggs into the strainer.
10. Repeat the plate-rinsing process as needed to get as many eggs as possible onto the strainer.
11. To initiate egg hatching, use bleach to remove the antennae-looking filaments (chorion) on the egg surface. Fill a container such as a 100 mm diameter Petri dish with undiluted bleach and another with $H_2O$.
12. Submerge the strainer containing the eggs in the bleach. Place this container under the dissecting microscope to see the chorion.
13. While looking through the dissecting microscope, gently swirl the strainer and use a paintbrush to break apart clumps.
14. When the eggs lose the chorion (i.e., dechorionated), remove the strainer from bleach and submerge in water (*see* **Note 9**).
15. Thoroughly rinse eggs with water using squirt bottle.
16. Transfer the strainer to clean container, such as a 100 mm diameter Petri dish, filled with water.
17. Scoop eggs from the strainer with a paintbrush and place on 1% agar (*see* **Note 10**).
18. Prepare a 17% (1x) yeast suspension in sterile water.

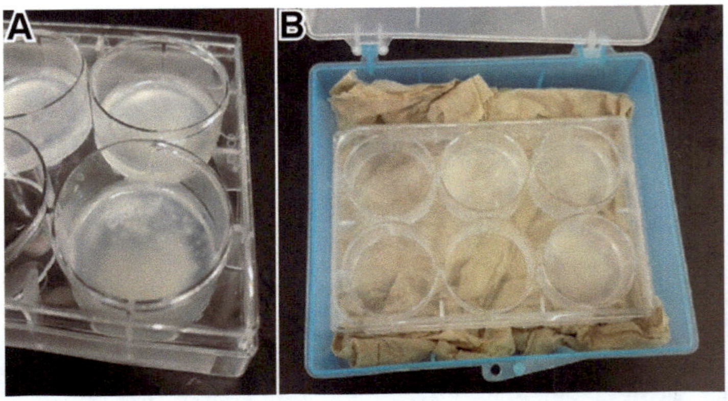


**Fig. 2** Example of a typical infection of larva. (**a**) Infection plate with 1% agar, bacteria/yeast suspension added; (**b**) simple humidified box made from pipette tip box and damp paper towels

19. Add approximately 50–100 μL of the yeast suspension to each well. The liquid should cover the surface of the well, but not pool (Fig. 2a).
20. Seal the plate with breathable sealing membrane to prevent larvae from escaping. Place plate in a humidified box and incubate at 22–26 °C overnight (Fig. 2b).
21. Inoculate 3–5 mL of HIB with desired *Y. pestis* pCD1$^{-}$ strains and incubate 18–24 h at 26 °C, with gentle aeration (~125 rpm).
22. Prepare a 34% (2×) yeast suspension in sterile water.
23. Measure $OD_{600}$ of *Y. pestis* cultures and calculate predicted colony forming units per milliliter (CFU/mL) (*see* **Note 11**).
24. Mix bacterial culture with yeast to get final predicted concentration of $1 \times 10^8$ CFU/mL. Prepare a control sample that is not infected by mixing an equivalent volume of HIB with 34% yeast.
25. Infect larvae by pipetting 100 μL of bacteria/yeast mixture into each well. Gently rock or rotate the plate to cover the surface of each well.
26. Seal wells with breathable membrane and incubate in humidified box at 26 °C. Larvae will pupate after 3 days.
27. Verify the concentration of the inoculum by serial dilution and plating on *Yersinia* selective agar (YSA).
28. Monitor the larvae once daily to ensure sufficient water and food in the plate. Throughout the infection process, keep the larvae fed and hydrated by adding more 17% yeast suspension and/or water as needed (*see* **Note 12**).
29. Depending on the experimental question, larvae may be harvested at a particular time point or once daily for up to 3 days (*see* **Note 13**). To harvest larvae, use a very fine soft-bristled paintbrush to scoop them out one at a time and transfer to a cell strainer. Multiple larvae harvested from the same well can be combined in the cell strainer.
30. Place 100% ethanol in a container such as a 60 mm or 100 mm diameter Petri dish; fill a second container with water.
31. After harvesting, rinse larvae well by gently immersing the strainers in 100% ethanol for 5–10 s, followed by water for 5–10 s.
32. Transfer individual larvae to microcentrifuge tubes (one/tube) containing 60 μL PBS.
33. Homogenize larvae using a microfuge pestle. Enumerate *Y. pestis* by serial dilution and plating on YSA (*see* Fig. 3 for typical infection results).

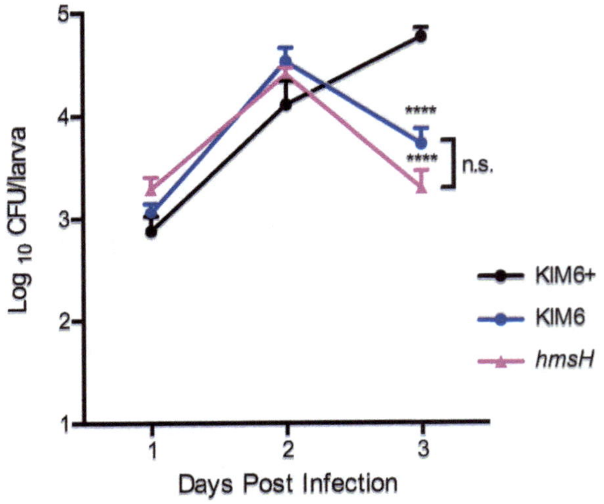


**Fig. 3** Example of a typical *Y. pestis* infection of *D. melanogaster* larvae. Larvae were infected and bacterial survival was evaluated. Infections were performed in triplicate ($n = 15$ per strain per day). Means and standard error are shown. ANOVA with Bonferroni multiple comparisons post-test was performed to make comparisons between strains for each day. $^{****}P < 0.0001$ compared to KIM6+, n.s. = no significance between mutants (reproduced from ref. 6)

### 3.3 Growing and Maintaining Schneider 2 (S2) Cells

1. Schneider 2 (S2) cells are grown at room temperature in T-25 cell culture flasks containing Schneider's media supplemented with 10% FBS and pen/strep. Cells should be split 1:5 when they reach confluency and can be passaged multiple times.
2. To generate frozen stocks, grow cells to high density. Centrifuge then remove growth media. Suspend cells in freezing media (75% FBS, 10% DMSO, 0.3% glucose, 5% RPMI or DMEM) (*see* **Note 14**).
3. To use for infection, remove 1 or 2 vials of S2 cells from liquid nitrogen storage, and thaw (*see* **Note 15**).
4. Transfer S2 cells to 5–6 mL Schneider's *Drosophila* medium supplemented with 10% FBS and pen/strep in a 15 mL tube.
5. Incubate for 30 min at room temperature.
6. Centrifuge at 500–1000 × *g* for 5 min.
7. Discard medium and replace with 5–6 mL Schneider's *Drosophila* medium.
8. Transfer cells in 25 cm$^2$ tissue culture flasks, incubate at 22–24 °C until cells are 80–90% confluent.
9. Remove the S2 cells from the old flask by gently scraping with a cell scraper or by firmly tapping the flask on your hand a few times.
10. Add 3–5 mL of cells to 10–15 mL Schneider's *Drosophila* medium and transfer to 75 cm$^2$ tissue culture flask (*see* **Note 16**).

### 3.4 *Y. pestis* Infection of S2 Cells

1. Streak bacterial strains for isolation on HIA-Congo red. Incubate at 26 °C for 2–3 days.
2. Inoculate an isolated colony into 3–5 mL HIB. Grow 18–24 h at 26 °C with gentle aeration (e.g., 120 rpm). Harvest stationary phase bacteria.
3. To prepare S2 cells for infection, remove cells from the culture flask.
4. Wash 1× in Schneider's media with 10% FBS (no antibiotics). Resuspend in 10 mL Schneider's media with 10% FBS.
5. Add Trypan blue to an equal volume of cells. Count the number of unstained live cells in 10 μL on a hemocytometer (*see* **Note 17**). Calculate the concentration of live cells per mL.
6. Seed 6-well plates at a density of $1 \times 10^6$ cells (*see* **Note 18**) in a total volume of 2 mL Schneider's media with 10% FBS (no antibiotics).
7. Incubate the plate at room temperature for 30 min to allow the cells to adhere to the wells.
8. Measure $OD_{600}$ of the bacterial culture. Calculate the expected bacterial concentration (*see* **Note 11**).
9. Centrifuge bacteria at 2000 × *g* for 2–5 min, resuspend pellet in Schneider's media with 10% FBS.
10. Determine the amount of bacteria to add for the desired multiplicity of infection (MOI) (*see* **Note 19**). Control wells should be given the same volume of Schneider's media with 10% FBS.
11. To initiate the infection, centrifuge plates at 500–1000 × *g* for 5 min.
12. Serial dilute and plate the inoculum on HIA-congo red to determine actual infection dose.
13. Incubate the infected cells at room temperature for the desired length of time (*see* **Note 20**), then process for appropriate assay (e.g., harvest RNA, fix for microscopy, or measure cytotoxicity or phagocytosis). Figure 4 shows scanning electron micrographs of S2 cells infected with *Y. pestis* KIM6- or CO92pCD1- after 1 h incubation at room temperature.

## 4 Notes

1. To reduce safety and security concerns, all of these protocols are performed in *Y. pestis* strains that lack the virulence plasmid pCD1, which encodes for the type III secretion system. These strains are avirulent in mammals and are classified as exempt

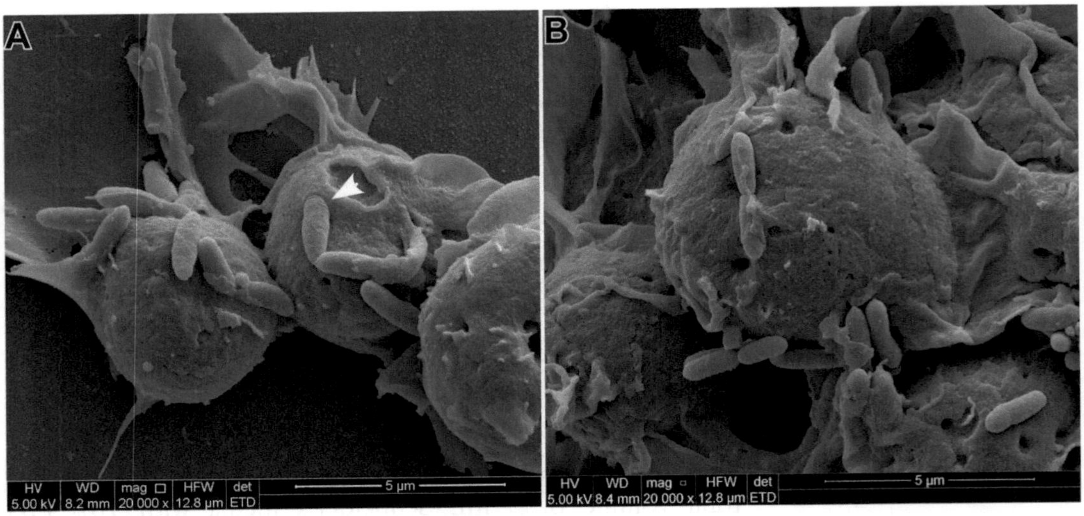


**Fig. 4** *Yersinia pestis* infection of *Drosophila* S2 cells. *Drosophila* S2 cells were infected at a multiplicity of infection (MOI) of 20 with *Y. pestis* (**a**) KIM6- or (**b**) CO92pCD1- (*pgm+*) *Y. pestis*. Cells were harvested and processed for scanning electron microscopy after 1 h of infection at room temperature. Dual-beam FIB SEM was used to acquire images showing adherent *Y. pestis*, primarily present as single cells or in pairs, the characteristic safety pin appearance. Also frequently observed were bacteria that appeared to be entering cells ((**a**), right cell)

from the select agent list, and therefore do not require an approved select agent registration for use.

2. The number of flies needed is dependent on the number of infections. About 50 flies should lay enough eggs to analyze two conditions (e.g., comparison of wild type and mutant bacterium).
3. Egg collection chambers are commercially available.
4. Pen/strep is commercially available at a 1000× concentrated stock.
5. *Drosophila* can be obtained through stock centers such as the Bloomington Stock Center. The Bloomington Stock Center also provides many protocols and media recipes for maintaining laboratory-reared *Drosophila*.
6. The frequency with which the fly strains should be passaged is affected by incubation temperature and fly genotype, but should generally not exceed 30 days between passages. Strains can be passaged for many years. Growth chambers should be cleaned with disinfectants such as bleach, ethanol, or soap to reduce the outgrowth of fungal and bacterial contaminants.
7. Heat-killed yeast is made by pouring yeast into a 125 mL flask so that it is a thin layer covering the bottom of the flask. Microwave on high power for 10 s, three times.

8. There are many available online resources, such as YouTube videos, for training in *Drosophila* egg collection as well as dechorionation.
9. This should take only take about 1–2 min and can be easily monitored under an inverted microscope.
10. Pour molten 1% agar into microtiter dishes shortly before use for maximal hydration of the agar. Up to 25 larvae per well for 12-well microtiter dishes; up to 100 larvae per well for 6-well microtiter dishes.
11. Predicted CFU/mL is calculated from empirically determined conversion factors. For each *Y. pestis* strain, determine the number of colony forming units in a liquid culture of $OD_{600}$ of 1. This can be used as a conversion factor to determine the predicted CFU of any given optical density value.
12. The amount of additional yeast/water mixture needed will depend on the evaporation in the chamber. It is important to keep the larvae hydrated with enough liquid to cover the surface of the well without forming standing pools.
13. The larva will live for 4 days before pupating, and over this time period, wild type *Y. pestis* pCD1- will grow at an exponential rate. We typically recover up to $10^5$ CFU/larva from a wild type infection.
14. There are many recipes for freezing media for S2 cells. S2 cells are sensitive to DMSO which may slow the initial recovery of cells from frozen stocks.
15. S2 cells grow very slowly when seeded at low density, so it is important to freeze cells at high concentration. Conversely, S2 cells grow very rapidly when density is high and daily observations may be needed to prevent overgrowth. Slow-thaw methods have given us better recovery of cells from frozen stocks. In addition, recovery can be improved if the cells are placed in the media before thawing is complete.
16. One 75 $cm^2$ flask of S2 cells should provide enough cells for one 12-well plate.
17. Cells should be counted after 1–2 min in Trypan blue. Overstaining will result in a large number of false positives.
18. Depending on the experiment, the size of the plates and the amount of media will vary. We typically use 2 mL for a 12-well microtiter dish and 5 mL for a 6-well microtiter dish.
19. S2 cells should be plated or seeded at high density for infections. For example, seed $1 \times 10^6$ cells per well of a 6-well microtiter dish or $5 \times 10^5$ cells per well of a 12-well microtiter dish. High MOIs (greater than 20) result in increased lysis of S2 cells.

20. The infection is incubated at room temperature in a closed container. The amount of incubation time is dependent on the downstream application. S2 cells are professional phagocytes.

## Acknowledgement

This work was supported by NIH/NIAID PHS award 1R21AI101242 to MMM and DMA. Special thanks to the University of Missouri Electron Microscopy Core and David Stella for the SEM imaging.

### References

1. Lorange E, Race B, Sebbane F, Hinnebusch B (2005) Poor vector competence of fleas and the evolution of hypervirulence in *Yersinia pestis*. J Infect Dis 191:1907–1912
2. Krasnov B (2008) Functional and evolutionary ecology of fleas: a model for ecological parasitology. Cambridge University Press, Cambridge
3. Mouse Genome Sequencing C (2002) Initial sequencing and comparative analysis of the mouse genome. Nature 420:520
4. Rennoll S, Rennoll-Bankert K, Guillotte M, Lehman S, Driscoll T, Beier-Sexton M, Rahman M, Gillespie JA, Azad AF (2017) The cat flea (*Ctenocephalides felis*) immune deficiency signaling pathway regulates *Rickettsia typhi* infection. Infect Immun 86(1): e00562–e00517
5. Earl S, Rogers M, Keen J, Bland D, Houppert A, Miller C, Temple T, Anderson D, Marketon M (2015) Resistance to innate immunity contributes to colonization of the insect Gut by *Yersinia pestis*. PLoS One 10(7):e0133318
6. Walker K, Maltez V, Hall J, Vitko N, Miller V (2013) A phenotype at last: essential role for the *Yersinia enterocolitica* Ysa type III secretion system in a *Drosophila melanogaster* S2 cell model. Infect Immun 81(7):2478–2487
7. Schneider I (1972) Cell lines derived from late embryonic stages of *Drosophila melanogaster*. J Embryol Exp Morphol 27(2):353–365
8. Myllymäki H, Rämet M (2013) Transcription factor zfh1 downregulates *Drosophila Imd* pathway. Dev Comp Immunol 39(3):188–197
9. Dimarcq J, Imler J, Lanot R, Ezekowitz R, Hoffmann J, Janeway C, Lagueux M (1997) Treatment of l(2)mbn *Drosophila* tumorous blood cells with the steroid hormone ecdysone amplifies the inducibility of antimicrobial peptide gene expression. Insect Biochem Mol Biol 27(10):877–886
10. Pearson A, Baksa K, Rämet M, Protas M, McKee M, Brown D, Ezekowitz R (2003) Identification of cytoskeletal regulatory proteins required for efficient phagocytosis in *Drosophila*. Microbes Infect 5 (10):815–824
11. Zhang B, Xu Z, Zhang Y, Shao X, Xu X, Cheng J, Li Z (2015) Fipronil induces apoptosis through caspase-dependent mitochondrial pathways in *Drosophila* S2 cells. Pestic Biochem Physiol 119:81–89
12. Surgalla M, Beesley E (1969) Congo red-agar plating medium for detecting pigmentation in *Pasteurella pestis*. Appl Microbiol 18 (5):834–837
13. Wu A, Lin G, Zhao F (2015) Data from: Transcriptome resources for the oriental rat flea and primary plague vector, *Xenopsylla cheopis* in Genomic Resources Notes Accepted 1 August 2015 to 31 September 2015. Dryad Data Repository. https://doi.org/10.5061/dryad.kq201s
14. Coordinators NR (2016) Database resources of the National Center for Biotechnology Information. Nucleic Acids Res 44(Database issue):D7–D19

# Part IV

## *Yersinia* Interaction with Immune Cells and Immune Signaling

# Chapter 13

# Intracellular Assays to Monitor Survival and Growth of *Yersinia pestis* Within Macrophages

**Amanda R. Pulsifer, Tiva T. VanCleave, and Matthew B. Lawrenz**

## Abstract

*Yersinia pestis* is able to survive and replicate within macrophages, while also being able to live in the extracellular milieu of the host. Assays that facilitate better understanding of how *Y. pestis* survives intracellularly and subverts normal host antimicrobial defenses require the ability to monitor intracellular *Y. pestis* survival and replication. In this chapter three different assays for monitoring intracellular survival and replication will be described, along with the formulas and methods to quantify and present the acquired data. These assays are fundamental to answering a multitude of questions pertaining to which bacterial factors are important for intracellular survival. Additionally, these assays can be used, with modifications, for other intracellular pathogens of interest. The first assay discussed will be the conventional bacterial enumeration assay, which quantifies bacterial numbers directly through a classic colony forming units (CFU) assay. Quantifying bacterial burden through CFU determination allows for differentiation between intracellular/cell-associated bacteria and extracellular bacteria. However, CFU determination is laborious, does not allow for direct kinetic monitoring of bacterial growth, and is difficult to adapt to high throughput assays. Bioluminescence bioreporters that use luciferase to monitor bacterial numbers allow for simple, plate reader-based, real-time kinetic monitoring of bacterial growth that is amendable to high throughput techniques. Finally, we will describe live cell microscopy using fluorescent bioreporters, which allows for monitoring of bacterial replication in individual cells and the possibility to visualize interactions between bacterial and host proteins during intracellular infection.

**Key words** Intracellular bacteria, Macrophages, Plague, Live cell microscopy, Bioluminescence

## 1 Introduction

*Yersinia pestis* is a facultative intracellular pathogen that can survive within human and laboratory mice macrophages and, to a lesser degree, neutrophils [1, 2]. Several lines of evidence suggest that intracellular survival in macrophages directly contributes to *Y. pestis* virulence. Mutants attenuated for intracellular survival in macrophages take longer, and/or require a higher inoculum, to cause a lethal infection [3, 4]. Ye et al. also demonstrated that depletion of macrophages and dendritic cells slows the dissemination of *Y. pestis* [5]. Finally, macrophages from species resistant to plague are

Viveka Vadyvaloo and Matthew B. Lawrenz (eds.), *Pathogenic Yersinia: Methods and Protocols*, Methods in Molecular Biology, vol. 2010, https://doi.org/10.1007/978-1-4939-9541-7_13, 

inherently more capable at killing intracellular *Y. pestis*, suggesting that host resistance to lethal infection is directly related to increased bacterial susceptibility to macrophage-mediated killing [6, 7]. These data support the continued need for in vitro studies to define the molecular mechanisms used by *Y. pestis* to subvert macrophage killing in susceptible hosts.

While *Y. pestis* can inhibit phagocytosis, the virulence factors involved in this process are not expressed in the flea vector [5]. Thus, upon transmission from the flea, *Y. pestis* is highly susceptible to phagocytosis [6]. Upon phagocytosis by macrophages, the bacteria appear to remodel the phagosome to resemble a recycling endosome in order to evade lysosomal fusion, resulting in the generation of a protective niche within the macrophage known as the *Yersinia* containing vacuole (YCV) [8, 9]. *Y. pestis* remains within this compartment throughout the entire course of intracellular infection, eventually replicating and killing the infected macrophage [10]. While the intracellular cycle of infection is beginning to be defined, the bacterial factors mediating YCV biogenesis still remain a mystery. Robust assays to monitor intracellular survival of *Y. pestis* will be instrumental in defining these molecular mechanisms.

In this chapter, we will focus on three assays to monitor intracellular *Y. pestis*. While each approach has its own advantages, they also have limitations, and often may be used in concert to study different aspects of *Y. pestis*–macrophage interactions. All three methods use similar protocols for seeding macrophages, with slight variations in cell number and adherence times based on the individual protocol, and preparation of bacterial cultures prior to macrophage infection. After addition of the bacteria to macrophages, the cells are briefly centrifuged to synchronize cell-bacteria interactions. As *Y. pestis* can replicate in many cell culture media, extracellular bacteria need to be removed using extensive washing and/or a gentamicin protection assay. Here, careful considerations should be made when designing the gentamicin protection assay to avoid erroneous conclusions based on unintended inhibition of intracellular bacteria by high concentrations of gentamicin [11]. We recommend empirical identification of the minimal gentamicin concentration and/or incubation times needed to kill extracellular bacteria prior to initiating intracellular survival assays, especially because different cell types have different propensities for gentamicin uptake. After completion of the gentamicin protection assay, the three methods diverge in sample processing and evaluation technique.

The first method that we will describe is a conventional bacterial enumeration assay using serial dilution and plating on agar plates. Bacterial enumeration of intracellular colony forming units (CFU) is the traditional gold standard for monitoring intracellular survival. However, this method is laborious, requires a large number of samples for time course analyses, does not allow for monitoring of the same sample over multiple time points, and is not suitable for high throughput biology. Recent advances in luminescent and fluorescent bioreporters have allowed for real-time and single cell monitoring of intracellular *Y. pestis* [12–14]. As such, we will also describe a bioluminescent approach for monitoring intracellular *Y. pestis* survival in real-time that is amenable to high throughput screening approaches in tissue culture. This method requires considerably fewer manipulations to quantify intracellular bacterial numbers than conventional enumeration, and allows for kinetic analysis of *Y. pestis* growth in the same sample over multiple time points. However, it is critical that users validate the correlation between the luminescence produced by an individual bioreporter to bacterial numbers, and that this correlation remains consistent over the entire course of infection. Furthermore, considerations in plate layout and positioning of plates in $CO_2$ incubators need to be made during assay design to minimize potential "edge effects" in bioluminescence. Finally, we will describe an approach to monitor *Y. pestis* in individual cells using live cell confocal microscopy. As intracellular and extracellular growth can be differentiated using microscopy, this approach can be used to study early events during infection without the use of gentamicin. However, when intracellular *Y. pestis* begin to lyse macrophages, and bacteria grow extracellularly in the medium, later time points in the infection process can be obscured. For this third approach, we focus specifically on infection, image acquisition, and quantification of intracellular bacteria by microscopy, but this approach can also be applied to monitor protein localization/trafficking using fluorescent fusions to host proteins. Although this chapter will detail methods to monitor intracellular *Y. pestis* in macrophages, these methods can also be adapted to other cell types and bacteria, including other species of *Yersinia*, with modifications.

## 2 Materials

### 2.1 Cell Culture

1. Macrophages (e.g., primary macrophages or macrophage cell lines, such as RAW264.7) (*see* **Note 1**).
2. T-75 cell culture flask.
3. Sterile cell culture grade 1× Phosphate Buffered Saline (1× PBS).
4. DMEM + 10% FBS.
5. 0.5% Trypan Blue.
6. Silicone cell scraper.
7. Hemocytometer or automated cell counter.
8. Pipette (P2–20, P20–200 and P100–1000).
9. Sterile 14 ml disposable tubes.

### 2.2 Bacterial Culture

1. *Y. pestis* strain suitable for CFU/RLU/Live cell (*see* **Note 2**).
2. Brain Heart Infusion (BHI) broth.
3. Culture tubes.
4. 26 °C incubator with roller drum or shaker set to 200 RPM.
5. Spectrophotometer.
6. Cuvettes for spectrophotometer.
7. Sterile 1× PBS.
8. 10 mg/ml gentamicin sulfate.

### 2.3 Conventional Y. pestis Enumeration

1. Sterile 1× PBS.
2. Multichannel pipettes (P2–10 μl and P20–200 μl).
3. 96-Well clear U-bottom plate.
4. 1% Triton.
5. Brain Heart Infusion (BHI) agar plates.
6. Sterile $H_2O$.

### 2.4 Y. pestis Enumeration Using Luciferase Bioreporter

1. White 96-well microtiter F bottom plate.
2. Luminescent plate reader such as Synergy HT.

### 2.5 Y. pestis Enumeration Using Live Cell Confocal Microscopy

1. 35 mm glass bottom dish.
2. Centrifuge adaptor plate for 35 mm dish (*see* **Note 3** and Fig. 1).
3. FluoroBrite DMEM supplemented with 10% FBS and 4 mM glutamine (*see* **Note 4**).
4. Confocal microscope equipped with a live cell imaging chamber, atmosphere regulator, and image acquisition software.
5. Fiji software.

## 3 Methods

### 3.1 Macrophage Preparation

1. Gently rinse macrophages in a T75 flask (80–90% confluent) by pipetting the old medium over the adherent cells two times with a serological pipette. Remove the old medium containing non-adherent cells/dead cell debris and dispose of appropriately.
2. Add 10 ml of fresh 37 °C DMEM +10% FBS and gently scrape the flask with a silicone cell scraper to dislodge the attached live cells.
3. Quantify cell concentration using trypan blue and a hemocytometer or automated cell counter.

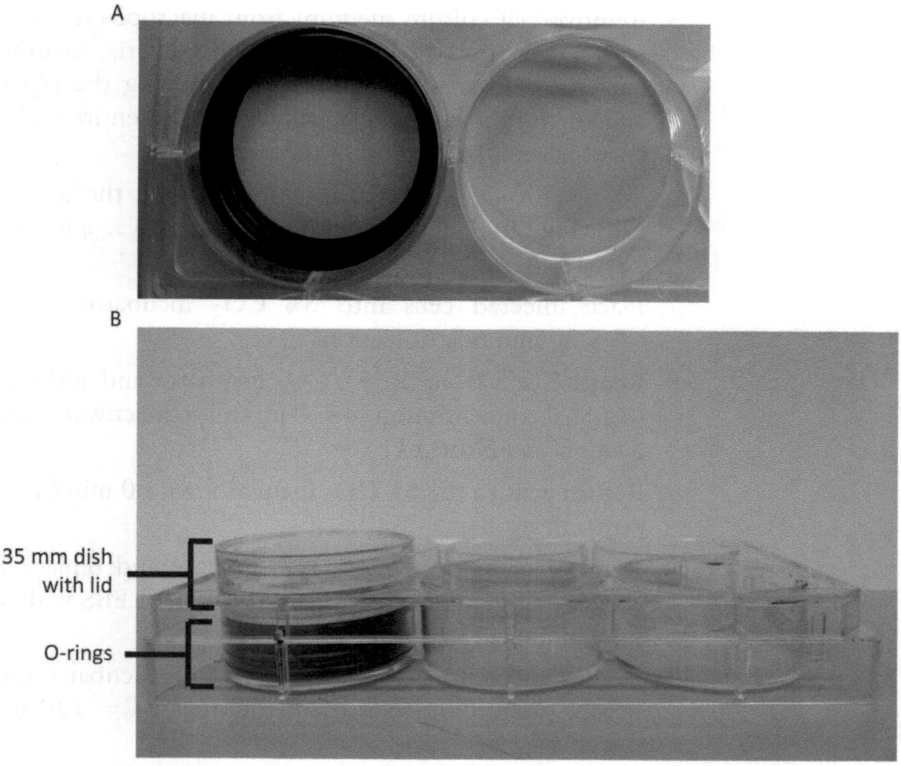


**Fig. 1** 35 mm dish adaptor plate configuration. (**a**) In order to centrifuge the 35 mm dish to synchronize the infection, an adaptor plate is used. The adaptor plate consists of a 6-well plate with four #45 rubber O-rings stacked on top of each other in one of the wells. (**b**) The glass bottom dish is inserted into the well containing the O-rings. The O-rings allow the lid for the dish to remain in place and prevent the dish from sinking to the bottom of the well, allowing for easy removal after centrifugation. Because the lid for the 35 mm dish remains in place, there is no need to use the lid for the 6-well plate during centrifugation

4. Dilute macrophages to desired concentration in 37 °C DMEM +10% FBS (*see* **Note 5**).
5. Pipette macrophages slowly into wells of the plate, moving the pipette back and forth in a sweeping motion to ensure an even distribution of cells over the entire well (*see* **Note 6**).
6. Allow the macrophages to attach to the plate for a minimum of 3 h before proceeding with *Y. pestis* infection (*see* **Note 7**).

### 3.2 Macrophage Infection with *Y. pestis*

1. 15–18 h prior to infection, inoculate 5 ml of BHI broth with *Y. pestis* and grow at 26 °C with aeration.
2. Dilute cultures 1:25 in 5 ml of fresh BHI broth and grow at 26 °C with aeration for 3 h (*see* **Note 8**).
3. Determine the bacterial concentration of the 3 h culture using a spectrophotometer (*see* **Note 9**).
4. Dilute bacteria to appropriate concentration in 37 °C DMEM +10% FBS based on the desired multiplicity of infection (MOI) (*see* **Note 10**).

5. Remove cell culture medium from macrophages and replace with DMEM +10% FBS containing bacteria. Gently pipette bacteria onto macrophages, slowly moving the pipette back and forth in a sweeping motion over the entire well/dish to ensure even distribution of the bacteria.
6. To synchronize bacterial interactions with the macrophages, centrifuge the microtiter plate/dish at 200 × *g* for 5 min (*see* **Note 3**).
7. Place infected cells into 5% $CO_2$ incubator for 15 min ($T = 20$ min post-infection).
8. Remove cells from the 5% $CO_2$ incubator and add gentamicin to a final concentration of 4–8 μg/ml to inactivate extracellular bacteria (*see* **Note 11**).
9. Return cells to the 5% $CO_2$ incubator for 60 min ($T = 80$ min post-infection).
10. Remove cells from 5% $CO_2$ incubator and replace medium containing gentamicin with DMEM +10% FBS with 1–2 μg/ml gentamicin (*see* **Note 12**).
11. Return infected macrophages to 5% $CO_2$ incubator for 40 min before proceeding to subsequent steps ($T = 120$ min post-infection).

### 3.3 Conventional Enumeration of Intracellular *Y. pestis*

For conventional enumeration, we typically use a 96-well microtiter plate. Individual wells of macrophages will need to be infected for each time point that will be analyzed (typically $n \geq 3$ per time point). The first time point harvested is at $T = 120$ min after initial gentamicin treatment. This time point represents the number of bacteria that are phagocytosed by the macrophage and serves as a reference point for subsequent time points used to monitor intracellular survival. The $T = 120$ min sample can also be compared to the inoculum to determine differences in *Y. pestis* invasion/phagocytosis.

1. Warm BHI agar plates by placing them in 37 °C incubator at least 30 min prior to harvesting of infected macrophages (*see* **Note 13**).
2. Prepare a serial dilution plate by adding 180 μl of sterile 1× PBS to Rows B–H of a sterile 96-well U bottom plate. You will need one column per macrophage sample (Fig. 2a).
3. At the desired time point, remove medium from infected macrophages and gently rinse cells two times with 0.5 ml of sterile 1× PBS by slowly dispensing onto the side of the well to limit detachment of cells.
4. Remove all of the 1× PBS and add 50 μl of 1% Triton X-100 to lyse infected macrophages. Incubate for 5 min at room temperature.

A

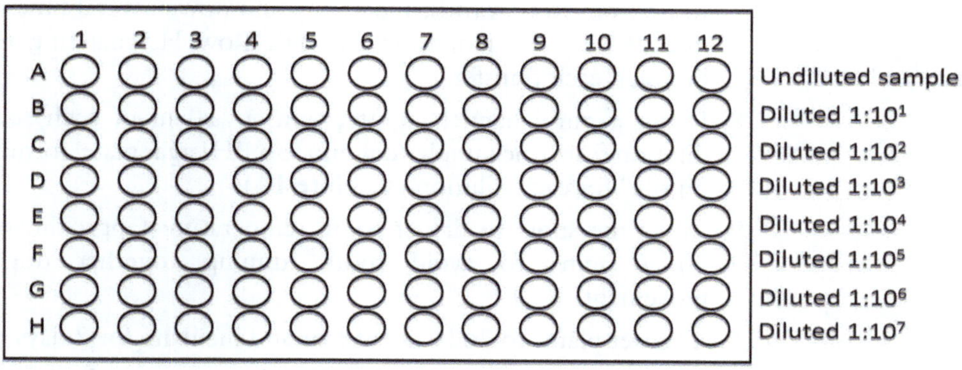


B

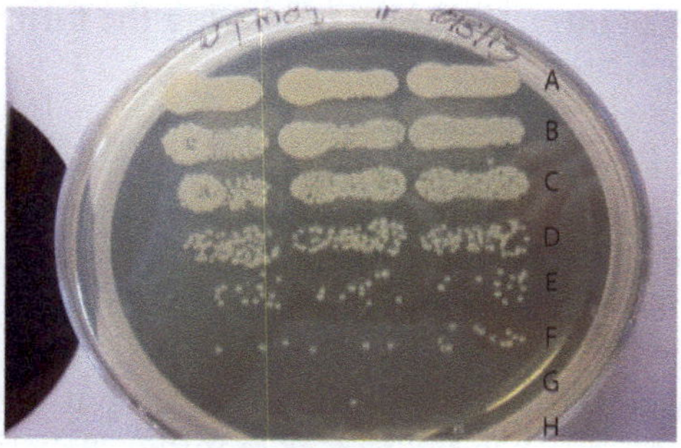


C

$$CFU/ml = \frac{Colonies \times \frac{1000}{Spot\,Volume}}{Dilution\,Factor}$$

$$CFU/well = \frac{CFU}{ml} \times Initial\,Well\,Volume$$

D

| Row Designation | Sample 1 | Sample 2 | Sample 3 | Dilution Factor | Spot Volume (μl) | Initial Well Volume (ml) | CFU/ml | CFU/Well |
|---|---|---|---|---|---|---|---|---|
| | | | | | | | | |
| A | Lawn | Lawn | Lawn | 1.00E+00 | 10 | 0.2 | | |
| B | Lawn | Lawn | Lawn | 1.00E-01 | 10 | 0.2 | | |
| C | Lawn | Lawn | Lawn | 1.00E-02 | 10 | 0.2 | | |
| D | TNTC | TNTC | TNTC | 1.00E-03 | 10 | 0.2 | | |
| E | 14 | 11 | 14 | 1.00E-04 | 10 | 0.2 | 1.29E+07 | 2.58E+06 |
| F | 4 | 2 | 12 | 1.00E-05 | 10 | 0.2 | 4.58E+07 | 9.16E+06 |
| G | | 1 | | 1.00E-06 | 10 | 0.2 | | |
| H | | | | 1.00E-07 | 10 | 0.2 | | |

**Fig. 2** CFU dilution plate and enumeration calculations. An agar plate with appropriate antibiotic selection for the strain is used to grow serial dilutions of intracellular bacteria from which the total number of intracellular bacteria can be calculated. (**a**) 96-well dilution plate schematic depicting the multichannel dilution scheme. (**b**) Agar plate with multichannel aliquots of serially diluted *Y. pestis* colonies to demonstrate spacing. (**c**) Calculations for CFU/ml and CFU/well to determine the total amount of *Y. pestis* from the sample. (**d**) Demonstration of calculation table for the agar plate shown in (**b**)

5. Add 150 μl of sterile 1× PBS and pipette multiple times to completely disrupt macrophage membranes and distribute bacteria.
6. Transfer macrophage lysate from 96-well microtiter plate to appropriate well in Row A of a 96-well plate (Fig. 2a). Repeat for all wells that were treated with Triton X-100.

7. Using a multichannel pipette, transfer 20 μl of macrophage lysate from Row A to B. Pipette up and down several times to mix. Repeat serial dilution through Row H, changing tips between each transfer.
8. Using a multichannel pipette, transfer 10 μl of each serial dilution from one sample/column to a BHI agar plate, forming "spots" for each dilution (*see* **Note 14**).
9. Allow the spots to dry at room temperature (approximately 20–30 min) to avoid spots running together during movement.
10. Transfer plates to 26 °C incubator and incubate for 2 days.
11. Remove plates from the incubator and enumerate colony forming units (CFU) for spots with individual colonies (Fig. 2b).
12. Calculate the number of intracellular bacteria using the formulas in Fig. 2c.

### 3.4 Kinetic Monitoring of *Y. pestis* Intracellular Growth Using a Luciferase Bioreporter

For kinetic monitoring of intracellular growth, we use a 96-well white microtiter plate to increase luminescence detection. Because the luminescence can be affected by "edge effects" we do not recommend using the outside wells. Typically we use the remaining columns per group, allowing for up to 10 groups with $n = 6$ technical replicates per plate. However, for high throughput biology applications, fewer replicates can be used as required. As written, this protocol outlines steps for a Biotek Synergy HT plate reader. However, other plate readers can be used, incorporating plate reader-specific modifications.

1. During Subheading 3.2, **step 6**, turn on the plate reader and adjust the chamber temperature to 37 °C (*see* **Note 15**).
2. At $T = 20$ min post-infection, prior to addition of gentamicin in Subheading 3.2, **step 8**, remove plate from incubator and transfer to plate reader.
3. Acquire luminescent data using the following settings: temperature 37 °C, 2 min delay, gain of 135, and integration time of 0.5 s (*see* **Note 16**).
4. Remove plate from plate reader and proceed with Subheading 3.2, **step 8**.
5. At $T = 120$ min post-infection, transfer plate to the plate reader and acquire luminescent data as described in Subheading 3.4, **step 3**.
6. Return plate to incubator.
7. To kinetically monitor intracellular bacterial numbers, repeat Subheading 3.4, **step 3** at desired time points, returning the plate to the incubator between time points.

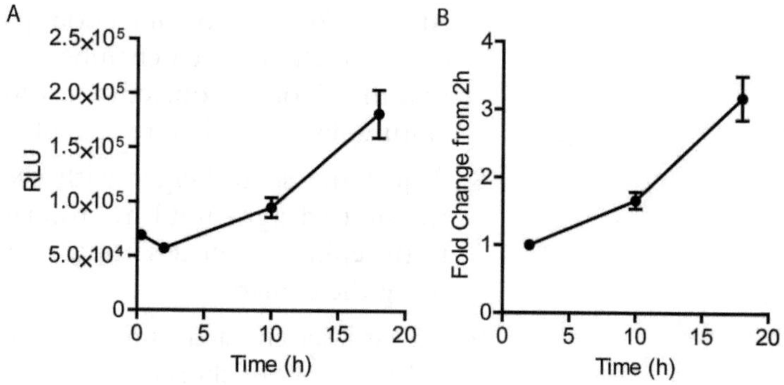


**Fig. 3** Bioluminescence monitoring of intracellular *Y. pestis* growth. RAW264.7 macrophages were infected with an MOI = 10 of *Y. pestis*. Extracellular bacteria were killed with gentamicin and intracellular bacterial luminescence was monitored for 18 h. (**a**) Raw relative light unit (RLU) values obtained from the Synergy HT plate reader. (**b**) Normalized RLU values where the time point of each well was divided by the 2 h time point to give the fold change over time

8. Data can be represented as relative light units or normalized to the $T = 120$ min time ([RLU at $T = X$ min]/[RLU at $T = 120$ min]) to represent survival over time (Fig. 3).

### 3.5 Live Cell Confocal Microscopy Imaging of *Y. pestis*

For live imaging we typically use a 35 mm glass bottom dish. However, other multiwell imaging systems can be used as long as they have glass bottoms compatible with the microscope being used. Oil or water immersion objectives may limit the number of wells that can be imaged for long periods of time, as movement of the objective to different locations over large areas will eventually result in loss of oil/water. Because of differences in microscope software, a generalized protocol is presented that needs to be modified depending on the equipment being used, and microscope and imaging chamber set up should be performed according to specific manufacturer guidelines.

1. During Subheading 3.2, **step 9**, set up the microscope and imaging chamber.
2. Turn on the microscope following the protocol recommended by the manufacturer.
3. Open $N_2$ flow to the compression table to stabilize the microscope.
4. At least 30 min prior to imaging, equilibrate the live cell imaging chamber by attaching and turning on heating elements and atmosphere regulators to produce a 37 °C, 5% $CO_2$ environment.
5. If using an oil immersion objective, add a drop of immersion oil to the objective (*see* **Note 17**).
6. Open the confocal microscopy software package for the microscope being used.

7. At $T = 120$ min post-infection, place the 35 mm glass bottom dish into the live cell chamber, adjust the objective height so that the oil on the objective just touches the bottom of the cell culture dish, and then re-assemble chamber.
8. Adjust the stage height with the fine adjustment using the transmitted light until macrophages come into focus, switch to the epifluorescence setting to visualize and focus on bacteria using the oculars.
9. Adjust laser and acquisition settings appropriate for the bacterial fluorophore chosen.
10. Make sure the Galvano imaging button is clicked.
11. Set the pinhole aperture to 1 AU on the bacterial fluorophore laser or the optimal setting to maximize picture quality.
12. Set the frame size to 1024 × 1024 pixels (*see* **Note 18**).
13. Adjust pixel dwell speed or frame rate to obtain the image resolution desired (e.g., a pixel dwell of 10.4 was used for Fig. 4).

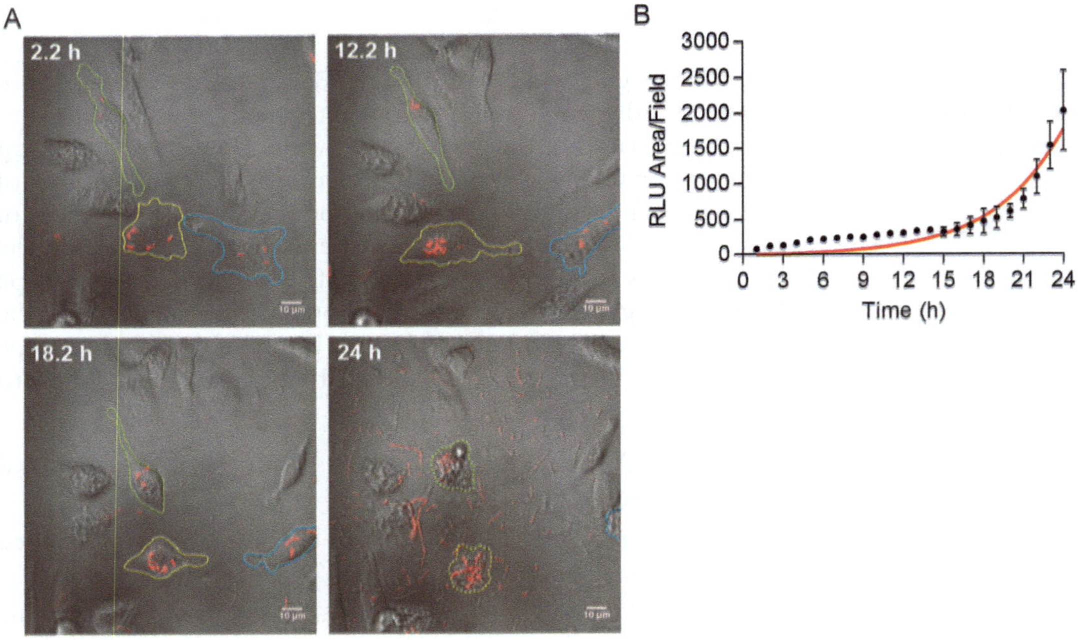


**Fig. 4** Enumeration of *Y. pestis* by live cell confocal microscopy. Peritoneal macrophages were infected with *Y. pestis* pgm$^{(-)}$ pCD1$^{(+)}$ expressing mCherry and live cell confocal microscopy was used to monitor the intracellular bacterial proliferation over a 24 h period. (**a**) Representative images of infected macrophages at indicated time points. Individual macrophages tracked through the entire course of the experiment are outlined in different colors. Lysed macrophages at 24 h are highlighted by dotted lines. Bacteria are red. (**b**) Bacterial numbers as a function of total fluorescent (RLU) area per field were determined every 1 h for 24 h. Each point represents the mean ± S.D. calculated from four imaged fields at each time point (an average of 16 infected cells per field). The red line represents the line of best fit. (Figure modified from VanCleave TT, Pulsifer AP, Connor MG, and Lawrenz MB (2017). Front. Cell. Infect. Microbiol., doi:https:/doi.org/10.3389/fcimb.2017.00505)

14. Increase averaging number as desired to enhance picture quality, but this will also increase the time required to capture each frame (averaging of 2 to 4 is usually sufficient, unless very high resolution images are desired).
15. Adjust the DIC settings so that the macrophage membranes are easily discernable.
16. Adjust the laser power and voltage for each of the laser lines so that fluorescence signal is visible but not producing saturated pixels (*see* **Note 19**).
17. Set up the acquisition settings as desired for *Z* stacks, Time series, or Multi-point (*XY*) images (*see* **Notes 20** and **21**).
18. Set up the time interval between acquisitions and the number of loops to perform (*see* **Note 22**).
19. Once all parameters are set, activate focus drift minimizing software (e.g., Perfect Focus) to limit sample drift over the course of the experiment.
20. Set the file to auto-save in a designated location.
21. With all the parameters set, click the "Run or Start Experiment" button.

### 3.6 Enumeration of *Y. pestis* from Live Cell Imaging Data

Here we describe using Fiji to quantify individual bacteria and overall fluorescent signal as a function of bacterial proliferation. However, multiple software packages are available that can be used for quantifying bacterial proliferation once images have been acquired.

1. Open the microscope extension file acquired containing the live cell image(s).
2. Convert images into 8 bit by selecting Image > Type > 8 bit.
3. Separate the channels by selecting Image > Color > Split Channels.
4. Threshold the bacterial channel in order to select only pixels with bacterial signal and to remove any background signal by selecting Image > Adjust > Threshold. Move the top slider to the point where only bacterial signal is highlighted and click apply. The bottom slider should not need to be adjusted. In general the auto adjust button is sufficient.
5. A pop-up window will appear with a method and background choice. Default can be used with a dark background. Calculated threshold for each image will be checked and the remaining will be unchecked. Click Ok.
6. A black and white image of the bacteria will now be displayed; this can be used for the next step.

7. To quantify the fluorescence signal, click Analyze > Analyze Particles. Many different parameters will be displayed that can be altered depending on need. In general, the size and circularity can be left in the default setting. For the "Show" option, choose outlines so that you can verify the program has correctly identified all of the bacteria. Leave the boxes checked and click Ok.
8. A window that says "Summary" will pop-up that contains all of the counts for the multiple time points that were acquired. The data can be transferred to Excel and parsed into the time points of interest such as 1 h, 2 h, 3 h, etc. Count, Total Area, Average Size, % Area and Mean for each time point will be displayed. The count and total area give the total number of bacteria and total fluorescent signal, respectively, and can be used to track bacterial proliferation (Fig. 4).

## 4 Notes

1. Primary macrophages can be isolated from animals or humans using standard protocols described elsewhere [15]. Macrophage cell lines can be maintained in cell culture using standard practices. For either, cells should be harvested, differentiated, and maintained without the addition of antibiotics, as residual antibiotics can remain in the cells for extended periods of time and inhibit intracellular *Y. pestis* survival.
2. Fully virulent or attenuated *Y. pestis* strains can be used to infect macrophages. Fully virulent strains require BSL-3 facilities and Select Agent approval. All manipulations of fully virulent or *pgm* negative bacteria should be performed within a biosafety cabinet (BSC) to minimize risk of aerosol exposure. Integration of *lux* genes or genes encoding fluorescent proteins into the chromosome or on shuttle plasmids have been developed as effective bioreporters for *Y. pestis* [12–14].
3. In order to centrifuge the 35 mm dish, four stacked #45 O-rings are inserted into a well of a 6-well tissue culture plate. The 35 mm dish is then inserted into the well. The O-rings allow the lid for the 35 mm dish to remain in place and prevent the dish from sinking to the bottom of the well, allowing for easy removal after centrifugation. As configured, the lid for the 6-well plate will not fit on the adaptor with the 35 mm dish inserted. However, because the lid for the 35 mm dish is in place, the 6-well lid is not required for preventing contamination of the culture. Place the adaptor into a swinging bucket type rotor configured for microtiter plates and lock the biosafety lid in place. Use a second 6-well dish (adding water to the empty wells for balance) to balance. Centrifuge at $200 \times g$

for 5 min at room temperature. Transfer the centrifuge bucket to a biosafety cabinet and remove biosafety lid and adaptor to eliminate exposure to potential aerosols generated during centrifugation.

4. FluoroBrite™ DMEM is formulated to have 90% lower background fluorescence than standard phenol red-free DMEM, providing a better signal-to-noise ratio during live cell imaging.
5. The number of cells used per well is assay dependent and varies primarily based on the well size used. For 24-well plates used for conventional enumeration, we typically use $1.5 \times 10^5$ cells/well. For 96-well white plates used for bioluminescent monitoring, we typically use at $1 \times 10^5$ cells/well. For 35 mm dishes used for live imaging, we typically use $5 \times 10^5$ cells/plate. These numbers can be decreased if longer incubation times of dividing cells (e.g., macrophage cell lines that divide) are used prior to infection with *Y. pestis.*
6. If the cells are all dispensed onto the center of the dish, this can result in the macrophages being tightly grouped together in the center of the well. This can inhibit visualizing individual cells during confocal microscopy, and potentially the multiplicity of infection, as bacteria can be distributed to areas without cells, limiting interactions during infection steps.
7. Depending on the cell type used, the incubation time can be extended for greater than 3 h. For terminally differentiated macrophages derived from the peritoneal cavity, it is recommended to use the cells as soon as possible, as we have observed decreased viability after the first 24 h post-isolation. Viability continues to decrease each day out to 72 h post-isolation.
8. Growth at 26 °C mimics bacterial growth in the flea and limits expression of the Ysc type three secretion system (T3SS). Under circumstances where expression of the T3SS is required, supplement the BHI medium with 2.5 mM $CaCl_2$ and incubate the 1:25 diluted culture at 26 °C for 1 h, then shift to 37 °C for an additional 2 h. The T3SS can be further induced by adding 20 mM $MgCl_2$ and 20 mM sodium oxalate instead of $CaCl_2$. Induction of the T3SS will significantly decrease phagocytosis by the macrophages and increase T3SS-mediated macrophage death.
9. In our hands, an $OD_{600} = 1.0$ is equivalent to $3 \times 10^8$ CFU. However, bacterial growth conditions may change this $CFU/OD_{600}$ ratio. Therefore, this conversion factor may need to be determined empirically for different growth conditions and spectrophotometers.
10. Typically we use an MOI of 10 bacteria per macrophage, except for live cell imaging, in which we use an MOI of 3. We have observed that higher MOIs can increase the rate at which

macrophages are lysed by *Y. pestis*, resulting in earlier release of intracellular bacteria into the medium. *Y. pestis* tends to quickly aggregate and settle in liquid medium, and it is imperative to pipette and/or vortex cultures well during serial dilutions and prior to addition to cells. Serial dilutions of inoculum should be plated on BHI agar for bacterial enumeration to confirm MOI.

11. While gentamicin is actively excluded from cells, some gentamicin will accumulate within the macrophages. Depending on the concentration used, the antibiotic can artificially inhibit intracellular survival of *Y. pestis* [11]. In our experience, primary cells acquire/accumulate gentamicin intracellularly more readily than macrophage cell lines [11]. Therefore, we typically use a final concentration of 8 μg/ml of gentamicin for RAW264.7 cells and 4 μg/ml for primary macrophages. Higher concentrations or longer incubation periods with gentamicin will impact the intracellular survival of *Y. pestis* [11].
12. For experiments longer than 8 h, we recommend maintaining infected RAW264.7 macrophages with 2 μg/ml gentamicin and primary macrophages with 1 μg/ml gentamicin. For live cell confocal imaging, cells do not need to be maintained in gentamicin, but if prevention of extracellular replication is desired it can be added.
13. Warming agar plates at 37 °C removes any condensation that has accumulated on the agar during storage that could cause spots to run together on the plate. Warming also decreases the amount of time it takes for the spots to absorb into the agar, which is required prior to transfer of plates to the incubator, and limits spots running together.
14. All serial dilutions from three samples can be spotted on a 100 mm BHI agar plate. To aid in enumeration, dilutions can be dispersed by slightly "dragging" the spot over a 0.5 in. area, or tilting the plate in one direction after dispensing the dilutions, to cause the spot to spread out as depicted (Fig. 2b).
15. Warming the plate reader prior to imaging minimizes temperature shifts to the cells. Typically, we allow the plate reader to warm to 37 °C and stabilize for 20 min prior to the first read.
16. The settings used in this protocol are specific for the *Yersinia pestis* $\text{Lux}_{\text{ptolC}}$ bioreporter [12]. Gain and integration time may need to be modified for other bioreporters.
17. For monitoring intracellular survival, we typically use a 40× or 60× objective, which allow for easy resolution of changes in the number of intracellular bacteria, but also provide an adequate number of infected and uninfected macrophages per field imaged. However, higher magnification objectives can be used if better resolution of individual bacteria/*Yersinia* containing vacuoles is required.

18. The pixel density for the acquired images can be changed if desired; however, the 1024 × 1024 setting is generally used as a means to balance between large file size and image quality.
19. It is essential to have the appropriate settings for visualizing fluorescent signal. If the signal is too weak then some bacteria might not be registered during quantification. If the signal is too strong, individual bacteria will begin to merge together, making it difficult to discern the accurate number of bacteria per field. Additionally, when automating quantification, measuring the overall fluorescent signal can be skewed by saturated signal.
20. Managing *Z* stacks, multipoint, and time lapse to all work together during one experiment is often difficult to achieve. Generally if the cell density is good, a *Z* stack or multipoint combined with time lapse acquisition is often sufficient. However, if all three are required, separate functions may have to be integrated into a multi-step script.
21. When utilizing the *Z* stack function, you will have to identify the top and bottom of the cell and define the number of slices per stack. In our experience, slices of 0.3–0.5 μM are sufficient to monitor bacterial replication.
22. We typically acquire an image every 20 min for 24 h unless *Z* stack images are acquired. For *Z* stacks, it is very important to consider the amount of time that the cells and bacteria will be exposed to the laser. If a high resolution, time-consuming *Z* stack with multiple small slices is required, the frequency of *Z* stacks should be limited to one per hour (or less) with a standard confocal microscope. Otherwise, extended exposure times to the laser may lead to artificial bacteria and/or macrophage killing.

## Acknowledgments

The authors acknowledge the James Graham Brown Cancer Center Confocal Imaging Core and Center for Predictive Medicine for Biodefense and Emerging Infectious Diseases for resources and technical support. Funding support was from NIAID (AI119557, AI097608) and NIGMS (GM125504) to M.B.L.

### References

1. Straley SC, Harmon PA (1984) Growth in mouse peritoneal macrophages of *Yersinia pestis* lacking established virulence determinants. Infect Immun 45(3):649–654
2. Spinner JL, Winfree S, Starr T, Shannon JG, Nair V, Steele-Mortimer O, Hinnebusch BJ (2014) *Yersinia pestis* survival and replication within human neutrophil phagosomes and uptake of infected neutrophils by macrophages. J Leukoc Biol 95:389–398
3. Oyston PCF, Dorrell N, Williams K, Li S-R, Green M, Titball RW, Wren BW (2000) The

response regulator PhoP is important for survival under conditions of macrophage-induced stress and virulence in *Yersinia pestis*. Infect Immun 68(6):3419–3425
4. Bozue J, Mou S, Moody KL, Cote CK, Trevino S, Fritz D, Worsham P (2011) The role of the phoPQ operon in the pathogenesis of the fully virulent CO92 strain of *Yersinia pestis* and the IP32953 strain of *Yersinia pseudotuberculosis*. Microb Pathog 50:314–321
5. Ye Z, Kerschen EJ, Cohen DA, Kaplan AM, van Rooijen N, Straley SC (2009) Gr1+ cells control growth of YopM-negative *Yersinia pestis* during systemic plague. Infect Immun 77 (9):3791–3806
6. Ponnusamy D, Clinkenbeard KD (2012) *Yersinia pestis* intracellular parasitism of macrophages from hosts exhibiting high and low severity of plague. PLoS One 7(7):e42211
7. Pachulec E, Abdelwahed Bagga RB, Chevallier L, O'Donnell H, Guillas C, Jaubert J, Montagutelli X, Carniel E, Demeure CE (2017) Enhanced macrophage M1 polarization and resistance to apoptosis enable resistance to plague. J Infect Dis 216:761–770
8. Connor MG, Pulsifer AR, Chung D, Rouchka EC, Ceresa BK, Lawrenz MB (2018) *Yersinia pestis* targets the host endosome recycling pathway during the biogenesis of the *Yersinia*-containing vacuole to avoid killing by macrophages. MBio 9(1):e01800-17
9. Pujol C, Klein KA, Romanov GA, Palmer LE, Cirota C, Zhao Z, Bliska JB (2009) *Yersinia pestis* can reside in autophagosomes and avoid xenophagy in murine macrophages by preventing vacuole acidification. Infect Immun 77 (6):2251–2261
10. Pujol C, Bliska JB (2003) The ability to replicate in macrophages is conserved between *Yersinia pestis* and *Yersinia pseudotuberculosis*. Infect Immun 71(10):5892–5899
11. VanCleave TT, Pulsifer AR, Connor MG, Lawrenz MB (2017) Impact of gentamicin concentration and exposure time on intracellular *Yersinia pestis*. Front Cell Infect Microbiol 7:505
12. Sun Y, Connor MG, Pennington JM, Lawrenz MB (2012) Development of bioluminescent bioreporters for in vitro and in vivo tracking of *Yersinia pestis*. PLoS One 7(10):e47123
13. Nham T, Filali S, Danne C, Derbise A, Carniel E (2012) Imaging of bubonic plague dynamics by in vivo tracking of bioluminescent *Yersinia pestis*. PLoS One 7(4):e34714
14. Sha J, Rosenzweig JA, Kirtley ML, van Lier CJ, Fitts EC, Kozlova EV, Erova TE, Tiner BL, Chopra AK (2013). A non-invasive in vivo imaging system to study dissemination of bioluminescent *Yersinia pestis* CO92 in a mouse model of pneumonic plague. Microb Pathog, 55, 39–50
15. Ray A, Dittel BN (2010) Isolation of mouse peritoneal cavity cells. J Vis Exp 35:1488. https://doi.org/10.3791/1488

# Chapter 14

# Human Neutrophil Isolation and Degranulation Responses to *Yersinia pestis* Infection

**Kara R. Eichelberger and William E. Goldman**

## Abstract

Neutrophils are the primary immune cell recruited to the site of bacterial infection, where they can rapidly deploy vesicles filled with various pro-inflammatory and anti-microbial proteins. This degranulation process, combined with oxidative and nitrosative mechanisms, is a major part of the initial host response to kill microorganisms. Neutrophils are one of the main cell types that interact with *Yersinia pestis* during infection, which is often lethal in the absence of prompt antibiotic treatment. Intradermal inoculation of *Y. pestis* results in bubonic plague, and inhalation of aerosolized droplets containing *Y. pestis* results in pneumonic plague. Although neutrophils are recruited to the site of inoculation during both bubonic and pneumonic plague, the neutrophils fail to clear *Y. pestis*, and, during pneumonic plague, contribute to the development of severe pneumonia. Subverting neutrophil responses is critical to the development of fulminant disease, yet the mechanisms by which *Y. pestis* impairs neutrophils are poorly understood. Cell culture models are important tools for studying *Y. pestis* interactions with immune cells. We describe a cell culture model for the infection of human neutrophils with *Y. pestis*. Neutrophils are isolated from human peripheral blood at high purity and subsequently infected with *Y. pestis*. We specifically focus on the application of this in vitro infection assay to the analysis of neutrophil degranulation responses.

**Key words** Neutrophil, *Yersinia pestis*, Degranulation, Granule exocytosis, Neutrophil isolation, Plague, Cell culture, Flow cytometry

## 1 Introduction

Neutrophils are a critical component of the normal innate immune response to bacterial infections. They are one of the first cell types recruited to the site of infection and can kill microorganisms by a variety of mechanisms, such as degranulation. Mature, circulating neutrophils contain numerous pre-formed vesicles (granules) that can discharge their protein contents rapidly in response to infection [1, 2]. Balancing the degree of neutrophil degranulation at the site of infection is critical, as granule proteins are essential for host defense but excessive degranulation can severely damage host tissues. Neutrophils contain four main sets of granules comprised of various enzymes and other proteins (Fig. 1). These granules can

Viveka Vadyvaloo and Matthew B. Lawrenz (eds.), *Pathogenic Yersinia: Methods and Protocols*, Methods in Molecular Biology, vol. 2010, https://doi.org/10.1007/978-1-4939-9541-7_14, © Springer Science+Business Media, LLC, part of Springer Nature 2019

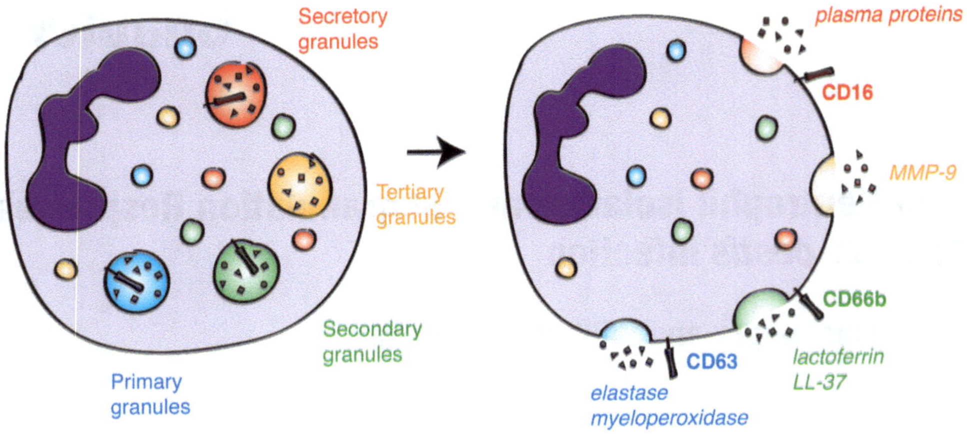


**Fig. 1** Neutrophils contain four main granule types that can be released to the extracellular environment in a process called degranulation. Secretory granules contain plasma proteins and replenish cell surface receptors such as CD16 after degranulation. Tertiary granules contain proteinases such as MMP-9 but are not defined by any unique cell surface receptors after degranulation. Secondary granules contain more pro-inflammatory proteins, such as lactoferrin and LL-37, and display CD66b on the neutrophil surface following degranulation. Primary granules contain the most pro-inflammatory and anti-microbial proteins, such as elastase and myeloperoxidase, and display CD63 on the cell surface after degranulation. A list of all the granule proteins and membrane markers contained within each of the granule types has been summarized by Cowland and Borregaard [24]

either fuse with the phagosome to kill intracellular microorganisms or fuse with the plasma membrane to kill extracellular microorganisms. Secretory granules are readily secreted and contain membrane proteins that promote neutrophil adhesion to the endothelium, whereas tertiary granules contain proteases such as MMP-9 that degrade the basement membrane to facilitate neutrophil extravasation from the bloodstream into the site of infection [3]. Secondary and primary granules, which contain anti-microbial and pro-inflammatory proteins, require the greatest stimulation for exocytosis, lead to the activation of monocytes and macrophages, and can cause neighboring cell and tissue damage [4–7].

Neutrophils are one of the main cell types that interact with *Yersinia pestis* during infection. *Y. pestis*, the causative agent of plague, is responsible for some of the deadliest disease outbreaks in recorded history. There are three main manifestations of plague that depend upon the route of inoculation. Bubonic plague results from intradermal inoculation, such as from a flea bite, pneumonic plague results from inhalation of aerosolized droplets containing *Y. pestis*, and septicemic plague results from direct intravenous inoculation [8]. Each *Y. pestis* infection is associated with high mortality rates in the absence of early antibiotic treatment. During infection, *Y. pestis* utilizes a type III secretion system (T3SS) to inject bacterial toxins, known as *Yersinia* Outer Proteins (Yops), directly into the cytoplasm of host cells [9]. The T3SS apparatus and associated Yop proteins are encoded on the pCD1 plasmid. Gene expression of the needle apparatus and Yop proteins is induced during growth at 37 °C, and Yops are secreted by the T3SS in a low calcium environment or following contact with host cells [9]. Neutrophils are a primary target of T3SS Yop injection [10, 11]. The *Y. pestis* T3SS leads to inhibited or altered neutrophilic activities, as Yops target host processes such as phagocytosis, cytokine secretion, and apoptosis [12–14]. Accordingly, neutrophils are impaired in responding to *Y. pestis* in both bubonic and pneumonic plague models. In bubonic plague, neutrophils are rapidly recruited to the site of *Y. pestis* inoculation in the dermis and control bacterial proliferation but fail to clear the bacteria from the tissue [15, 16]. During pneumonic plague, neutrophils are severely delayed in reaching the lungs and fail to clear *Y. pestis* or prevent bacterial dissemination to the bloodstream [17, 18]. Instead, neutrophils continue to accumulate in the airways and contribute to the development of pro-inflammatory lesions until the host ultimately succumbs to severe pneumonia [10, 19]. Understanding the mechanisms by which *Y. pestis* impairs or dysregulates neutrophil responses may reveal new host-oriented therapies for treating plague infections during later stages of disease.

Cell culture models have played a critical role in elucidating mechanisms of *Y. pestis* virulence. However, neutrophils are terminally differentiated and so cannot be maintained in cell culture, and they have a short life span (hours) following isolation [20]. Here, we describe the isolation and purification of human neutrophils from peripheral blood samples and subsequent infection of neutrophils with *Y. pestis* in vitro. We specifically focus on the application of this cell culture model to study neutrophil extracellular degranulation responses. Discharge of each granule type exposes specific markers on the surface of intact neutrophils, and we quantify relative levels of degranulation by measuring the fluorescent intensity of these markers using flow cytometry.

## 2 Materials

### 2.1 Bacterial Strains and Media

1. *Yersinia pestis* CO92. This strain is a Tier 1 Select Agent and requires the appropriate registration and approval with the Centers for Disease Control. Additionally, a Biosafety Level 3 (BSL-3) laboratory is required for working with this strain. If avirulent strains of *Yersinia pestis* are used (i.e. those that lack the *pgm* locus or the pCD1 plasmid), a Biosafety Level 2 (BSL-2) laboratory is needed.
2. Brain Heart Infusion Agar: 52 g/L, autoclaved 120 °C for 30 min and poured into petri dishes (*see* **Note 1**).
3. Brain Heart Infusion Broth: 37 g/L, heated until dissolved and filter sterilized.

### 2.2 Neutrophil Isolation from Human Peripheral Blood Samples

1. Venous blood collection tube with EDTA.
2. 3% Dextran/0.9% NaCl (for 200 mL): 6 g dextrose, 1.8 g sodium chloride in deionized water ($dH_2O$). Filter sterilized.
3. 50-mL conical tubes.
4. Serological pipets.
5. 0.9% NaCl (for 200 mL): 1.8 g sodium chloride in $dH_2O$.
6. 0.2% NaCl (for 200 mL): 0.4 g sodium chloride in $dH_2O$.
7. 1.6% NaCl (for 200 mL): 3.2 g sodium chloride in $dH_2O$.
8. Ficoll-Paque PLUS.
9. 25 G needle.
10. 10-mL Luer lock syringe.
11. Neutrophil stock medium: RPMI 1640 (without phenol red and HEPES, with L-glutamine) with 25 mM HEPES, pH 7.2, and filter sterilized (*see* **Note 2**).
12. Neutrophil working medium: neutrophil stock medium, 5% fetal bovine serum, 2.0 mM L-glutamine, and 2.5 mM $CaCl_2$ (*see* **Note 3**).
13. Fetal bovine serum (FBS).
14. Phosphate-buffered saline (PBS).
15. 24-well tissue culture plate.
16. Trypan blue (0.4% solution).
17. Hemocytometer slide and light microscope.

### 2.3 Neutrophil Infection with *Y. pestis*

1. PBS.
2. Spectrophotometer and cuvettes.
3. 37 °C incubator with nutating or rocking platform.

### 2.4 Cell Staining and Flow Cytometry Analysis

1. Flow buffer: 2% FBS in PBS. Make fresh before each use and store at 4 °C.
2. Antibodies: mouse anti-human CD63-PE-Cy7 (eBioscience), mouse anti-human CD66b-PE (eBioscience).
3. Annexin V Apoptosis Detection Kit FITC (eBioscience) (*see* **Note 4**).
4. Flow cytometer with blue laser (488 nm).

## 3 Methods

### 3.1 Bacterial Culture for Infection

1. Streak out *Yersinia pestis* strain onto a BHI agar plate and grow at 26 °C for 2 days.

2. Inoculate a 2-mL BHI broth culture with multiple colonies of *Y. pestis* from the agar plate. Incubate overnight at 26 °C in a roller drum.
3. Add 200 μL of the *Y. pestis* overnight culture into 10 mL BHI + 2.5 mM $CaCl_2$ in a 125-mL Erlenmeyer flask. Shake the culture at 37 °C at 250 rpm for 5 h (*see* **Note 5**).

### 3.2 Neutrophil Isolation from Human Peripheral Blood

Obtaining blood from human subjects requires institutional IRB approval. Only certified and trained individuals should perform the phlebotomy. Human blood should be treated as potentially infectious material and handled with appropriate biosafety measures.

1. Draw 30 mL of blood from a consenting donor into three EDTA venous blood collection tubes, with 10 mL of blood in each tube (*see* **Note 6**).
2. Pool the 30 mL of blood in a 50-mL conical tube (*see* **Note 7**).
3. Fill the remainder of the 50-mL conical tube with 3% dextran/0.9% NaCl solution, approximately 20 mL. Invert conical tube twice to mix.
4. Incubate at room temperature for 20 min. Two layers will form: a lighter top layer containing cells and plasma and a lower darker layer containing sedimented red blood cells [21] (*see* **Note 8**).
5. Use a 10-mL serological pipet to transfer the top cellular layer to a new conical. There will be 30–35 mL of this layer (*see* **Note 9**).
6. Centrifuge at 20 °C for 10 min at 250 × *g* with slow brake.
7. Aspirate the supernatant and resuspend the cell pellet in 25 mL of 0.9% NaCl.
8. Using a 10-mL Luer lock syringe with a 25 G needle, fill the syringe with 10 mL of Ficoll-Paque. Insert the syringe into the conical tube, placing the needle at the bottom of the tube. Slowly depress the syringe, adding the Ficoll-Paque beneath the blood resuspension. Two clear layers will form: the lower Ficoll layer and the upper layer containing the blood resuspension. Remove the syringe steadily from the liquid, taking care not to disturb the distinct interface between the two layers (*see* **Note 10**).
9. Centrifuge at 20 °C for 40 min at 400 × *g* with slow brake.
10. There will be four layers that form following the Ficoll-Paque density centrifugation. Neutrophils and remaining red blood cells are contained in the pellet at the bottom of the conical tube. Peripheral blood mononuclear cells (PBMCs) are contained in an interphase layer between the plasma (top layer) and the Ficoll (below PBMC layer). Remove the plasma,

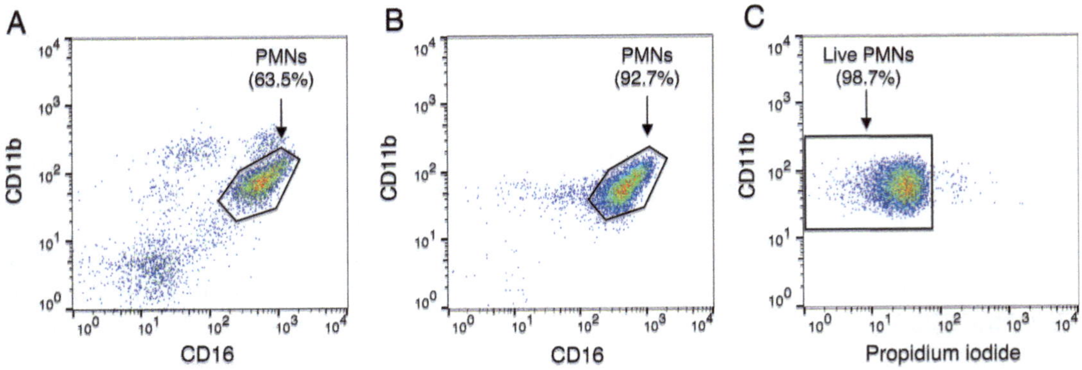


**Fig. 2** Neutrophils can be isolated with high purity and minimal cell death from human peripheral blood using Ficoll-Paque density centrifugation. (**a**) 10 mL of whole blood was stained with CD11b-FITC (eBioscience) and CD16-PE-Cy7 (eBioscience) following red blood cell sedimentation and lysis. Neutrophils (PMNs) are CD11b$^+$ and CD16$^+$, and PMNs make up approximately 60% of cells in whole blood. (**b**) Following isolation by Ficoll-Paque density centrifugation, neutrophil purity from whole blood can be increased to >90%. (**c**) 98% of purified neutrophils stain negative for live/dead discriminating dye propidium iodide (PI)

PBMC, and Ficoll layers, leaving behind the neutrophil/RBC pellet (*see* **Note 11**).

11. Resuspend the pellet in 20 mL of 0.2% NaCl solution, incubating for 30 s at room temperature. This will lyse red blood cells.
12. Add 20 mL of 1.6% NaCl solution to restore tonicity.
13. Centrifuge at 20 °C for 6 min at 250 × *g* with slow brake.
14. Aspirate the supernatant and repeat **steps 11** through **13** to lyse remaining red blood cells (*see* **Note 12**).
15. Resuspend the purified neutrophil pellet in 1 mL of PBS, and keep on ice.
16. Create a 1:100 dilution of neutrophils in 0.2% trypan blue diluted in PBS. Load 10 μL of the mixture onto a hemocytometer slide, and using a light microscope at 20× magnification, count the number of live neutrophils in each of the four quadrant counting grids (*see* **Note 13**). The average of this number is the average total number of neutrophils isolated, in millions. Typically, from 30 mL of blood we are able to isolate an average of 80 million neutrophils with >90% purity and >98% viability (Fig. 2).
17. Add the purified neutrophils to the needed volume of neutrophil working medium to yield a concentration of $2 \times 10^6$ neutrophils per 1 mL medium (*see* **Note 14**). Add 500 μL of neutrophils to each well to be used in the FBS-coated 24-well tissue culture plate (*see* **Note 15**). This yields a final concentration of $1 \times 10^6$ neutrophils per well.
18. Allow neutrophils to adhere to tissue culture wells at 37 °C for 30 min prior to infection with *Y. pestis* (*see* **Note 16**).

### 3.3 Infection of Neutrophils with Yersinia pestis

1. Prepare the *Yersinia* inoculum from the 10-mL culture grown at 37 °C. Take the $OD_{620}$ of each *Yersinia* culture to be used in the assay. 1 $OD_{620} = 1 \times 10^8$ CFU *Y. pestis.*
2. Centrifuge $5 \times 10^7$ CFU *Y. pestis* from the 10-mL culture at maximum speed (>16,000 × *g*) for 1 min. Remove the supernatant and resuspend the bacterial cell pellet in 1 mL of PBS. Keep the *Y. pestis* inoculum at 37 °C until infection.
3. To infect neutrophils, add 20 μL of inoculum ($1 \times 10^6$ CFU *Y. pestis*) to each of the designated wells. This yields an MOI of 1:1 (*see* **Note 17**). After adding one inoculum or infecting a group of at least three wells, swirl the contents of the 24-well plate to facilitate dispersion of the inoculum throughout the well.
4. Incubate plate for 1 h at 37 °C on a nutating or rotating platform (at slow speed) to mix bacteria and neutrophils throughout the infection (*see* **Note 18**).

### 3.4 Staining of Neutrophils for Flow Cytometry

1. After infection, collect the cells and bacteria into 1.7-mL Eppendorf tubes. Pipet the liquid up and down 2–3 times around the wells to fully dislodge adherent neutrophils (*see* **Note 19**).
2. Pool the wells designated for single-color controls into separate tubes for the single-color control samples (*see* **Note 20**).
3. Centrifuge the samples for 2 min at 1000 × *g*.
4. Aspirate the supernatant from each of the tubes, taking care not to disturb the neutrophil pellet.
5. Resuspend each of the pellets in 500 μL of cold flow buffer and centrifuge for 2 min at 1000 × *g*.
6. Aspirate the supernatant. Resuspend the neutrophil pellets in CD63-PE-Cy7 and CD66b-PE antibodies diluted in flow buffer for a total volume of 50 μL (*see* **Note 21**). Neutrophils that are releasing secondary granules will have increased fluorescent staining for CD66b, and neutrophils that are releasing primary granules will have increased fluorescent staining for CD63.
7. Incubate the samples on ice or at 4 °C for 30 min in the dark.
8. Add 500 μL of cold flow buffer to each sample and centrifuge for 2 min at 1000 × *g*.
9. Aspirate the supernatant. Resuspend the pellet in 500 μL of cold 1× Binding Buffer from the Annexin V Apoptosis Detection Kit.
10. Centrifuge samples for 2 min at 1000 × *g*.
11. Aspirate the supernatant and resuspend the samples in 100 μL 1× Binding Buffer containing Annexin V-FITC antibody

(*see* **Note 22**). Annexin V binds to phosphatidylserine, which is exposed on the surface of apoptotic neutrophils. Incubate at room temperature for 15 min. During this time, incubate the single color control tube for propidium iodide at 80 °C. This will kill enough neutrophils to serve as a positive control for the live/dead discrimination.

12. Add 500 μL of 1× Binding Buffer to each of the samples and centrifuge for 2 min at 1000 × *g*. Aspirate the supernatant.
13. Resuspend the samples in 300 μL 1× Binding Buffer containing propidium iodide (5 μL per 300 μL of 1× Binding Buffer). Any dead cells that have lost membrane integrity will take up the propidium iodide within 5 min and stain positive.
14. Acquire samples on a flow cytometer within 4 h (*see* **Note 23**).

### 3.5 Sample Analysis and Neutrophil Degranulation Quantification

Samples acquired from the flow cytometer can be analyzed using software such as FlowJo. Exemplary data analyzed using the gating scheme detailed below is shown in Fig. 3.

1. Plot the samples by forward scatter area and side scatter area and set a gate to exclude debris (*see* **Note 24**).
2. Plot the gated cells by propidium iodide (PI) and Annexin V-FITC. Live, non-apoptotic ("healthy") neutrophils will fall in the lower right corner of the plot. Early apoptotic neutrophils will be Annexin $V^{+}$ and $PI^{-}$, and necrotic or dead cells will be $PI^{+}$. Gate the healthy neutrophils.
3. Create two histogram plots from the healthy neutrophil gate: one with CD63-PE-Cy7 and one with CD66b-PE plotted against side scatter. Measure the mean fluorescence intensity (MFI) for each of the samples for both CD63 and CD66b. Overlay graphs can also be made from the histograms for visual comparison of the samples of interest. Calculate the average and standard deviation of each of the three replicate wells for a sample and graph. A higher MFI correlates with increased extracellular degranulation.

## 4 Notes

1. *Y. pestis* will not grow on Brain Heart Infusion medium mixed with agarose separately, thus it is best to use Brain Heart Infusion Agar to make plates.
2. When used under sterile conditions, neutrophil stock medium can be stored up to a year at 4 °C. It is important that the pH be buffered to 7.2. If the medium is too alkaline, the addition of calcium will cause calcium phosphate to precipitate out of solution, which will interfere with the assay.

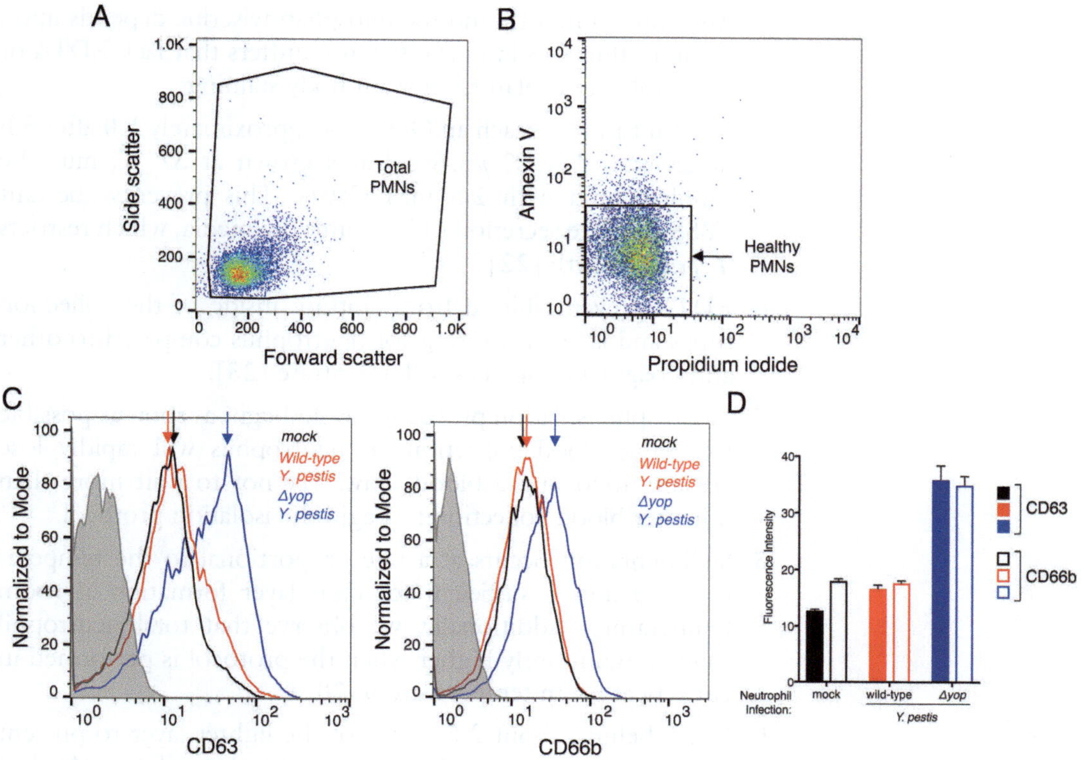


**Fig. 3** Quantification of human neutrophil extracellular degranulation following *Y. pestis* infection in vitro by flow cytometry. Exemplary data in this figure was acquired on a Millipore Guava easyCyte 6HT cytometer. (**a**) The sample is gated to exclude debris in a gate labeled "Total PMNs". (**b**) Events within the Total PMNs gate are further gated on Annexin V-FITC and propidium iodide levels. "Healthy PMNs" (non-apoptotic and viable) will fall in the lower left quadrant of the plot. This population is gated for analysis of granule marker surface expression [13]. (**c**) Two histogram plots can be generated from the Healthy PMNs gate, one for CD66b-PE (secondary granule marker levels) and one for CD63-PE-Cy7 (primary granule marker levels). Overlay plots can show the differences in granule marker expression for different treatment groups. Here we plot representative data comparing mock-infected neutrophils (black), neutrophils infected with wild-type *Y. pestis*, or neutrophils infected with *Δyop Y. pestis*, which is a strain that lacks all seven Yop effectors [9]. These differences are highlighted on the plots with an arrow pointed to the peak of marker levels for each of the three groups. Unstained neutrophils are shown in a shaded grey histogram. (**d**) The mean fluorescence intensity is measured for each of the samples. The average value for the mean fluorescence intensity for each replicate within a group is calculated and graphed for both CD63 and CD66b levels.

3. Neutrophil working medium is made fresh the day it is needed. We have tested both normal human serum and fetal bovine serum supplemented media and observed no difference in neutrophil degranulation responses dependent upon the species of serum used.
4. We find the kit works well in this protocol, but the individual components can also be purchased separately: Annexin V-FITC antibody, propidium iodide (7AAD may also be used), and a flow buffer that is free of calcium chelators. The interaction of

Annexin V antibody and the phosphatidylserine depends upon calcium, thus it is important to use buffers that lack EDTA or other calcium chelators after antibody staining.

5. This culture will reach an $OD_{620}$ of approximately 1.0 after 5 h of growth. Any *Y. pestis* cultures grown at 37 °C must be supplemented with 2.5 mM $CaCl_2$. This prevents aberrant T3SS-mediate secretion of Yops into the media, which restricts *Y. pestis* growth [22].
6. EDTA prevents blood from clotting inside of the collection tubes and is less activating for neutrophils compared to other anti-coagulants, such as sodium citrate [23].
7. Neutrophil isolation protocol should begin as soon as possible following blood collection, as neutrophils will rapidly lose viability following a blood draw. Try not to wait more than 2 h after blood collection to begin the isolation protocol.
8. Sedimentation occurs at a rate proportional to the temperature. 20 min is sufficient for clear layer formation at room temperature. Additionally, we observe that total neutrophil yield is significantly higher when the protocol is performed in entirety at room temperature or 20 °C.
9. Leave behind about 2.5–3 mL of the lighter layer to prevent accidentally removing some of the darker, lower layer. This lost volume will not reduce the number of neutrophils isolated with the protocol.
10. Inject the Ficoll at a rate of about 1 mL every 3 s, taking about 30 s to release all of the Ficoll. This rate will preserve the clear interface between the two layers. If the syringe is held upright and removed slowly and steadily after adding the Ficoll layer, the two layers will not be disturbed. It is critical to maintain two layers (one Ficoll layer and one cell resuspension layer) with a distinct interface for cell separation to occur.
11. To prevent contamination of the PBMC/monocyte layer with the neutrophil pellet during this step, begin aspiration with the pipet tip at the PBMC layer. We use a pipet tip attached to a vacuum line for this step. Move the pipet tip in concentric circles to aspirate the entire PBMC layer with the top layer, leaving the lower Ficoll layer to keep the PBMCs from coming in contact with the neutrophil pellet. Once we can no longer see PBMCs on top of the Ficoll layer, we aspirate the remaining liquid above the neutrophil pellet.
12. There will be some red blood cells that remain in the pellet, but these will not interfere with the neutrophil assay.
13. Live neutrophils will appear round and clear. Any cells that are dead will take up the Trypan blue dye and appear blue, and should be excluded from the total number of neutrophils.

We find that very few (<1%) neutrophils are dead following isolation.

14. When designing our assays, we use three wells for each strain to be tested. We include a mock-infected control group, a wild-type *Y. pestis* strain, and an avirulent *Y. pestis* strain (such as pCD1$^{-}$ *Y. pestis*) for comparison to all other strains tested.
15. The tissue culture plate can be coated with FBS by filling each well with 500 μL FBS and letting the plate incubate at 37 °C for 1 h. Following incubation, remove the FBS and wash the wells twice with PBS. Coating can be done during the dextran-red blood cell sedimentation step and the plate washed during the red blood cell lysis steps. The wells should not be allowed to dry completely before seeding the neutrophils.
16. When using BSL-3 strains of *Y. pestis* in our assays, we perform the neutrophil isolation protocol in a BSL-2 facility and transport the cells to BSL-3 for infection at this step.
17. If a different MOI is required, the bacterial CFU pelleted and resuspended in 1 mL can be adjusted. It is easier to adjust CFU of bacteria to change the MOI rather than the number of neutrophils.
18. Neutrophil degranulation is a rapid process, and phenotypes can be clearly detected after 1 h of infection.
19. This is best done holding the plate at a slight angle, aspirating the liquid that pools at the bottom of the well, and dispensing it at the top of the well to run down to the bottom again.
20. The single-color controls can be made by either pooling a small portion of each of the replicate wells used in the experiment or by pooling separate replicate wells designated for use as single-color controls, and then dispersing an equal amount of cells to each of the control tubes. It is best to have some portion of each strain or sample represented in the single-color controls.
21. Antibodies should be titrated to determine the appropriate concentration for use. To titrate antibodies, we begin with the recommended concentration on the product specifications and make four serial 1:2 dilutions. Six samples pooled from each of the test conditions are incubated with one of the antibody dilutions, plus one sample left unstained. Follow the protocol for antibody staining and washing. Acquire the samples on the flow cytometer, and determine the lowest concentration at which you can detect a clear difference in fluorescence intensity as compared to the unstained sample.
22. The Annexin V-FITC antibody can be titrated as well, as described in **Note 21**.
23. If samples are to be acquired on a flow cytometer located outside of the BSL-3 lab, the *Y. pestis* in the sample must first

be inactivated and verified to have all *Y. pestis* killed within the sample before removal from the BSL-3 lab, as per Federal Select Agent Program requirements. We use 2% formalin to fix cells and inactivate bacteria, but each lab must verify this method of Select Agent inactivation before use. Additionally, if samples are being fixed, a fixable live/dead dye must be used instead of PI or 7AAD. Fixed samples can be acquired on a flow cytometer the next day rather than within 4 h.

24. You can also set a threshold gate to exclude debris. The sample shown in Fig. 3 was acquired with a threshold gate set at 500 for the side scatter parameter on a Millipore Guava easyCyte 6HT cytometer.

## References

1. Kolaczkowska E, Kubes P (2013) Neutrophil recruitment and function in health and inflammation. Nat Rev Immunol 13:159–175
2. Lacy P (2006) Mechanisms of degranulation in neutrophils. Allergy Asthma Clin Immunol 2:98–108
3. Sengeløv H, Follin P, Kjeldsen L, Lollike K, Dahlgren C, Borregaard N (1995) Mobilization of granules and secretory vesicles during in vivo exudation of human neutrophils. J Immunol 154:4157–4165
4. Soehnlein O, Lindbom L (2009) Neutrophil-derived azurocidin alarms the immune system. J Leukoc Biol 85:344–351
5. Pereira HA, Shafer WM, Pohl J, Martin LE, Spitznagel JK (1990) CAP37, a human neutrophil-derived chemotactic factor with monocyte specific activity. J Clin Invest 85:1468–1476
6. Sahoo M, Del Barrio L, Miller MA, Re F (2014) Neutrophil elastase causes tissue damage that decreases host tolerance to lung infection with *Burkholderia* species. PLoS Pathog 10:e1004327
7. Winterbourn CC, Kettle AJ, Hampton MB (2016) Reactive oxygen species and neutrophil function. Annu Rev Biochem 85:765–792
8. Perry RD, Fetherston JD (1997) *Yersinia pestis*--etiologic agent of plague. Clin Microbiol Rev 10:35–66
9. Viboud GI, Bliska JB (2005) *Yersinia* outer proteins: role in modulation of host cell signaling responses and pathogenesis. Annu Rev Microbiol 59:69–89
10. Pechous RD, Sivaraman V, Price PA, Stasulli NM, Goldman WE (2013) Early host cell targets of *Yersinia pestis* during primary pneumonic plague. PLoS Pathog 9:e1003679
11. Marketon MM, DePaolo RW, DeBord KL, Jabri B, Schneewind O (2005) Plague bacteria target immune cells during infection. Science 309:1739–1741
12. Spinner JL, Cundiff JA, Kobayashi SD (2008) *Yersinia pestis* type III secretion system-dependent inhibition of human polymorphonuclear leukocyte function. Infect Immun 76:3754–3760
13. Stasulli NM, Eichelberger KR, Price PA, Pechous RD, Montgomery SA, Parker JS, Goldman WE (2015) Spatially distinct neutrophil responses within the inflammatory lesions of pneumonic plague. MBio 6:e01530–e01515
14. Spinner JL, Hasenkrug AM, Shannon JG, Kobayashi SD, Hinnebusch BJ (2016) Role of the *Yersinia* YopJ protein in suppressing interleukin-8 secretion by human polymorphonuclear leukocytes. Microbes Infect 18:21–29
15. Gonzalez RJ, Lane MC, Wagner NJ, Weening EH, Miller VL (2015) Dissemination of a highly virulent pathogen: tracking the early events that define infection. PLoS Pathog 11: e1004587
16. Bosio CF, Jarrett CO, Gardner D, Hinnebusch BJ (2012) Kinetics of innate immune response to *Yersinia pestis* after intradermal infection in a mouse model. Infect Immun 80:4034–4045
17. Lathem WW, Crosby SD, Miller VL, Goldman WE (2005) Progression of primary pneumonic plague: a mouse model of infection, pathology, and bacterial transcriptional activity. Proc Natl Acad Sci U S A 102:17786–17791
18. Bubeck SS, Cantwell AM, Dube PH (2007) Delayed inflammatory response to primary pneumonic plague occurs in both outbred and inbred mice. Infect Immun 75:697–705
19. Pechous RD, Broberg CA, Stasulli NM, Miller VL, Goldman WE (2015) In vivo

transcriptional profiling of *Yersinia pestis* reveals a novel bacterial mediator of pulmonary inflammation. MBio 6:e02302–e14–11

20. Amulic B, Cazalet C, Hayes GL, Metzler KD, Zychlinsky A (2012) Neutrophil function: from mechanisms to disease. Annu Rev Immunol 30:459–489
21. Kuhns DB, Long Priel DA, Chu J, Zarember KA (2015) Isolation and functional analysis of human neutrophils. Curr Protoc Immunol 111:7.23.1–7.2316
22. Michiels T, Wattiau P, Brasseur R, Ruysschaert JM, Cornelis G (1990) Secretion of Yop proteins by *Yersiniae*. Infect Immun 58:2840–2849
23. Freitas M, Porto G, Lima JLFC, Fernandes E (2008) Isolation and activation of human neutrophils in vitro. The importance of the anticoagulant used during blood collection. Clin Biochem 41:570–575
24. Cowland JB, Borregaard N (2016) Granulopoiesis and granules of human neutrophils. Immunol Rev 273:11–28

# Chapter 15

# Measurement of *Yersinia* Translocon Pore Formation in Erythrocytes

**Tiago R. D. Costa, Monika K. Francis, Salah I. Farag, Tomas Edgren, and Matthew S. Francis**

## Abstract

Many Gram-negative pathogens produce a type III secretion system capable of intoxicating eukaryotic cells with immune-modulating effector proteins. Fundamental to this injection process is the prior secretion of two translocator proteins destined for injectisome translocon pore assembly within the host cell plasma membrane. It is through this pore that effectors are believed to travel to gain access to the host cell interior. *Yersinia* species especially pathogenic to humans and animals assemble this translocon pore utilizing two hydrophobic translocator proteins—YopB and YopD. Although a full molecular understanding of the biogenesis, function and regulation of this translocon pore and subsequent effector delivery into host cells remains elusive, some of what we know about these processes can be attributed to studies of bacterial infections of erythrocytes. Herein we describe the methodology of erythrocyte infections by *Yersinia*, and how analysis of the resultant contact-dependent hemolysis can serve as a relative measurement of YopB- and YopD-dependent translocon pore formation.

**Key words** Contact-dependent hemolysis, Type III translocon pore complex, Biogenesis, Function and regulation, Membrane integration, Effector recognition and intracellular delivery, Host immune response

## 1 Introduction

### 1.1 The Type III Secretion System

Type III secretion (T3S) is an effective means for many different Gram-negative bacteria to deliver proteins into diverse eukaryotic cell types of mammalian, insect, fish, and plant origin. Not only is T3S a common virulence mechanism of many harmful pathogens [1, 2], but it is also a useful device for bacteria to establish a symbiotic lifestyle or to survive in the environment by preventing predation by protozoa and amoebae [3], and for the biosynthesis of the flagella apparatus [4, 5].

Upon artificial induction, a T3S system (T3SS) can also secrete substrates into laboratory culture media. This secretion requires assembly of a complex of 20–25 proteins spanning both bacterial

Viveka Vadyvaloo and Matthew B. Lawrenz (eds.), *Pathogenic Yersinia: Methods and Protocols*, Methods in Molecular Biology, vol. 2010, https://doi.org/10.1007/978-1-4939-9541-7_15, © Springer Science+Business Media, LLC, part of Springer Nature 2019

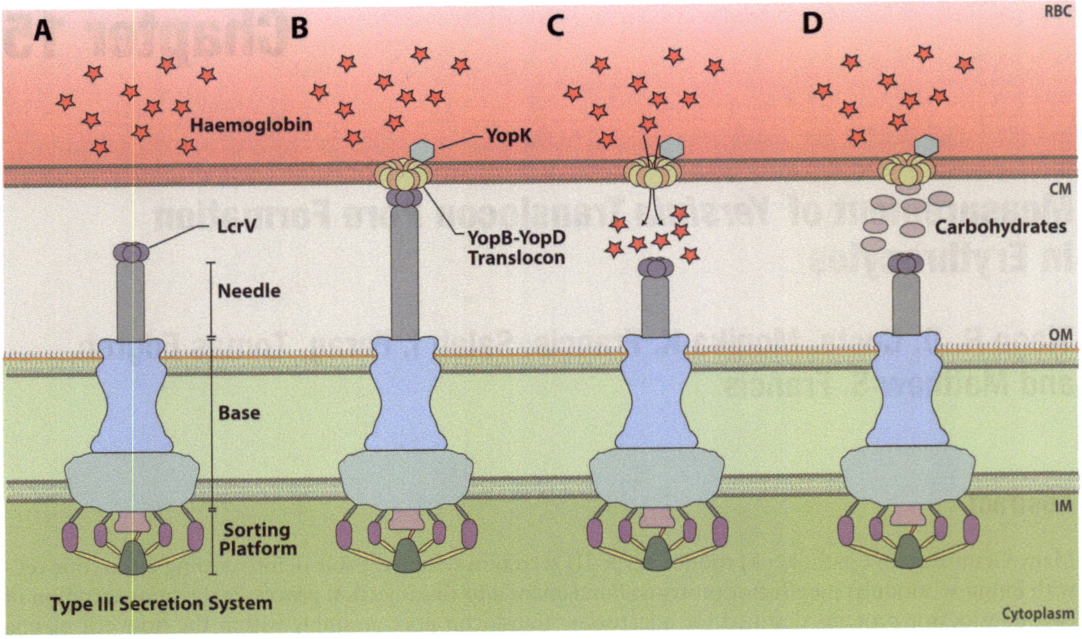


**Fig. 1** The *Yersinia* T3S translocon complex induces contact-dependent disruption of erythrocyte membranes. (**a**) A type III secretion system spans the entire bacterial envelope. The base component breaches the inner membrane (IM), periplasm and the outer membrane (OM). Assembled from the base is an extracellular needle appendage that extends out toward a host cell. Sitting atop this needle appendage is a needle-tip protein that in *Yersinia* is termed LcrV. Recognition of cytoplasmic-located substrates destined for type III secretion is performed by the sorting platform. Altogether, a fully assembled type III secretion system that is competent for substrate secretion consists of a minimum of 20 different proteins. With just a few exceptions, in *Yersinia* these are designated as Ysc (*Yersinia* secretion) proteins. (**b**) Upon contact with host cells, the *Yersinia* T3SS targets for secretion of the YopB and YopD proteins. With assistance from LcrV, these two proteins assemble a translocon pore in the host cell plasma membrane. Subsequent to this, Yop effector proteins are recognized for secretion and can transit through the membrane-inserted pore to gain entry inside the host cell. The process of effector delivery has been omitted from this figure. One of these internalized effectors is the YopK protein, which regulates both the biogenesis and function of the T3S translocon pore from inside the host cell. (**c**) As a model system for translocon pore biogenesis, *Yersinia*-infected sheep red blood cell (RBC) is used often. Presumably succeeding bacterial disconnect, infected RBC undergo osmotic lysis as a consequence of free bacterial induced pore formation in the plasma membrane. Hemoglobin released into the surrounding can be quantitated spectroscopically and these levels are a reflection of the degree of pore formation. (**d**) Pore size can be estimated by the addition of carbohydrates of various radii. A sufficiently sized carbohydrate can act as osmoprotectant by plugging the pore and preventing osmotic lysis and hemoglobin release

membranes (inner and outer), the peptidoglycan layer, and protruding out from the surface. The structure of this assembled T3S apparatus is likened to a hypodermic needle (termed the "needle complex") (Fig. 1a) [6–8]. It is assumed that unfolded protein cargo is secreted through the needle complex [9, 10]; a process energized by the combined effect of the proton motive force and ATP hydrolysis [11–14].

Protein cargo is divided into three secretion classes—"early" proteins that make up the protruding needle appendage (Fig. 1a), which is required for the subsequent secretion of "middle" proteins comprising the translocon complex and "late" proteins that possess host-modulating effector activity once targeted to the eukaryotic cell interior [15–17] (Fig. 1b). Many of the effector protein properties have been studied in detail in immune cells, providing molecular understanding of their immunomodulatory impact on host cell fate [18–20]. In contrast, knowledge of how the effectors are actually delivered across the eukaryotic cell membrane is still lacking.

### 1.2 The Translocon Pore Complex

The final step of the delivery of effectors into host cells is a function ascribed to the translocon complex positioned at the distal end of the needle (Fig. 1b). This complex integrates into the host cell plasma membrane to build-up a structure known as the injectisome translocon pore [21, 22]. It is assumed that formation of this membrane-spanning pore completes the entire T3SS assembly process and bridges the bacterium with the host cell.

A long-standing model of T3SS-dependent effector delivery is that effector proteins located in the bacterial cytoplasm are translocated directly into the host cell cytosol through the needle–translocon complex in a one-step process. However, recent data has emerged to suggest also an alternative two-step model in which proteins are first secreted to the bacterial surface, from where they can then be translocated into the host cell interior [23, 24]. Regardless of whether translocation occurs in one or two steps, it is likely that this process is closely coupled to an ability of the components to sense host cell contact and to sequester host cell proteins that contribute to pore assembly and activity [25–28]. Moreover, it has also been reported that the pore constituents are post-translationally modified [29], and a portion of them are also actively translocated that can have effector function inside host cells [15, 30, 31]. However, the significance of these observations is not yet clear, but it does indicate requirements for multiple contributors during translocon-pore function.

In human pathogenic *Yersinia* utilizing a plasmid encoded Ysc-Yop T3SS, the translocon complex formed in host cell membranes is composed of two proteins, YopB and YopD [31–36] (Fig. 1b). YopB and YopD were originally identified on the basis of their hydrophobic properties, which is a unique feature restricted to this "middle" class of type III secreted substrates [37]. In addition, the secreted proteins LcrV and YopK are required to regulate assembly and/or function of the translocon pore (Fig. 1b). LcrV localizes to the tip of the needle complex [38], where it serves as a pore-forming platform for the insertion of YopB and YopD into the host cell plasma membrane [39–41] and contributes also to the rapid intracellular targeting of a sub-class of Yop effectors that have responsibility for inhibiting phagocytic activity of host cells [42]. The biological role of YopK is less defined, but from inside host cells it contributes to limiting excessive pore formation, while also contributing to the correct spatial localization of injected Yop effectors [43–46]. It is likely that aspects of these functions involve a direct interaction with YopB and/or YopD [38, 43, 45, 47, 48].

It has been demonstrated using tissue culture monolayer infection assays that human pathogenic *Yersinia* can utilize their Ysc-Yop T3SS to inject effector proteins into diverse primary cells and transformed cell lines [49]. However, using mouse infections to model the systemic disease process, *Yersinia* appeared to preferentially target dendritic cells, macrophages, and neutrophils cells upon colonization of mesenteric lymph nodes and the spleen [50–52]. By extrapolation it is assumed that these immune cells are the actual targets of *Yersinia* during a natural human infection.

### 1.3 Erythrocytes (Red Blood Cells)

Erythrocytes (red blood cells—RBCs) function as the principle oxygen transporter in the body of diverse species. Hemoglobin is the main constituent of the cytoplasmic contents, and this conjugated metalloprotein is the actual oxygen carrier molecule [53, 54]. They are a fascinatingly unique eukaryotic cell for their final maturation is marked by the extraordinary loss of their nucleus and internal organelle structures [54, 55]. This lack of internal structures permits the cells to adopt a flattened disk-like shape that increases surface to volume ratio, and to achieve greater structural deformability to squeeze through small capillaries; two features that promote oxygen transport capacity [56, 57]. This is aided by a flexible membrane that is composed of an asymmetrical lipid bilayer, which anchors a number of proteins that make the membrane permeable to certain solutes, as well as providing connection points to the underlying shape-determining spectrin-rich cytoskeleton [56–58].

In healthy individuals, the RBC has a lifespan of about 4 months, with approximately 1% of the population replaced every day. This age restriction is forced because mature RBCs have lost transcriptional and translational capacity, which prevents the renewal of aged or damaged molecular infrastructure [59]. The monitoring of aged red blood cells occurs through the recognition of changes in the asymmetrical distribution of components within the lipid bilayer. The mechanisms of clearance follow two established pathways. The first is present in all cells and is dependent of IgG recognition of a senescence-specific antigen at the cell surface that appears following a conformational change in an integral membrane protein ("Band 3") [60]. IgG-opsonized RBCs are removed by the phagocytic activity of macrophages. The second is a specific erythrocyte programmed cell death pathway known as eryptosis [61]. This is a process by which phosphatidylserine appears at the outer surface of the RBC membrane, facilitating direct recognition and engulfment by CD36-positive macrophages.

### 1.4 Measurement of Translocon Pores Formed in Erythrocyte Membranes

Regardless of the in vivo significance, RBCs have for many years been the model cellular system of choice for studying structure-–function relationships of diverse trans-membrane penetrating pore-forming toxins [62, 63]. Pore insertion in the RBC plasma membrane leads to water influx, cell swelling, and subsequent osmolysis. This culminates in the release of hemoglobin, and this serves as a readily quantifiable marker permitting a relative measurement of pore formation by correlating to the level of osmolysis. The popularity of this cell model stems from the fact that it is a technically simple means to evaluate bacterial pore-forming activity in an easily accessible population of comparatively stable primary cells of reduced physiological complexity.

Within this backdrop, routine use of erythrocytes to investigate the molecular basis of bacterial pore formation by the T3SS translocon was triggered by a few pioneering studies [35, 64–66]. Since these early days, the *Yersinia*-infected (contact-dependent) RBC hemolysis assay has become a vital tool to study YopB- and YopD-dependent pore formation (Fig. 1c). The extent of RBC plasma membrane disruption, osmolysis, and the subsequent release of hemoglobin has most often been proportional to the level of pores formed in infected erythrocyte cell membranes as confirmed by effector translocation into target cells [32, 34, 35, 44, 67–73]. (*See* Subheading 1.5 for a cautionary tale.) With these observations, it is believed that pore formation in erythrocytes is a sound reflection of pore formation in vivo during *Yersinia* colonization. Hence, this chapter will describe the materials and methodology of contact-dependent RBC hemolysis as derived from many years of focused research aimed at prying apart the inner workings of the *Yersinia* T3S translocon pore complex. Experiments built around this classical hemoglobin release assay have provided the basis for understanding the minimal molecular components required for a functional translocon complex, and allowed investigations into the molecular interactions among these structural components. They have also enabled a better understanding of translocon pore assembly by revealing clues about translocator protein integration into host cell membranes and the size of this assembled pore.

### 1.5 Cautionary Tales and an Eye to the Future

Undoubtedly the contact-dependent RBC hemolysis assay has benefitted understanding of the dynamics of translocon pore biogenesis, regulation and function. As a cautionary note however, the user must be aware of weaknesses in the assay, even though the reasons for these weaknesses are poorly understood. In a recent study by Ekestubbe and others [42], *Yersinia* bacteria producing a well-defined stable mutant variant of the needle tip protein, LcrV, failed to induce contact-dependent hemolytic activity and did not insert YopB or YopD into erythrocyte membranes. This was a surprise given that these same mutant bacteria delivered the immunomodulatory YopE and YopH effectors into nucleated host cells through a process that was still typically dependent upon the YopB and YopD translocator proteins [42]. As this implied the unlikely scenario that effector translocation could proceed in the absence of pore assembly (at least as defined by the absence of contact-dependent osmolysis of infected RBCs), the authors investigated the extent to which these mutant bacteria could integrate YopB and YopD into the plasma membrane of nucleated cells (specifically the HeLa epithelial cell line was used). In these challenging membrane-enrichment experiments, indeed at least YopD (and by inference probably YopB) was actually membrane integrated [42]. On the basis of these findings in infected nucleated cells, some experimental situations indicate that a correlation is lacking between pore formation in erythrocytes and in vivo T3SS activity. Indeed, in some recent cases involving studies of defined YopD mutants, an imperfect correlation between contact-dependent hemolytic activity of RBCs and effector translocation into nucleated cells was observed [47, 69].

Highlighting these observations is not intended to deter the reader from using a pore formation assay based on the erythrocyte infection model; after all it still remains an intensely used assay shedding light on translocon pore assembly in host cell plasma membranes. So why do discrepancies occur between infections of anucleated and nucleated cells? Answers probably lie in the very reason that makes erythrocytes a powerful model in the first place. Since they have no nucleus, mature RBCs do not undergo transcription or translation, and they exhibit a restricted metabolic capacity [54, 55]. Yet, we know that translocon pores formed in nucleated immune cells trigger aggressive innate immune responses that depend upon transcriptional, translational, and metabolic activity [67, 72, 74, 75]. Indeed, it is this responsiveness from the host cell that necessitates subsequent intracellular delivery of effector proteins with immune-modulatory capacity, i.e., to quench active host responsiveness to plasma membrane integration of the T3SS translocon [48, 75, 76]. In contrast, there is very little support for the T3SS-dependent delivery of effectors into RBCs [77], which is consistent with RBCs not being a natural target of T3SSs in vivo [50–52]. It is also notable that apart from intrinsic membrane shedding during the aging process [78], RBCs lack sophisticated membrane repair systems that are functional in nucleated cells [79]. Hence, a translocon pore formed in the plasma membrane of nucleated cells will then be subjected to an array of cellular physiological responses that are either lacking or compromised in the affected anucleated RBC. These physiological differences in cell model systems can easily account for discrepancies in readouts observed from structure and function studies of the translocon pore.

Thus, we now recommend that future studies should utilize contact-dependent hemolysis of erythrocytes in combination with other systems to dissect the highly dynamic process of translocon pore biogenesis and function. In parallel, the reader is encouraged to explore the measurement of pore formation in infected nucleated cells, either by enrichment of infected membranes [42, 45, 47, 69, 80], or via studying uptake of the ethidium homodimer-2 [81, 82], release of 2′,7′-Bis(2-carboxyethyl)-5(6)-carboxyfluorescein acetoxymethyl ester [32, 33, 67, 81, 83, 84] or release of lactate dehydrogenase [33, 73, 76, 83–85]. Critically, employing parallel anucleated and nucleated cellular-based infection models to study translocon pore formation has the benefit of improving scientific robustness of the experimental findings.

It is also significant that we now appear to have the possibility to purify biologically active translocon components to homogeneity [30]. This should open up avenues to expand the experimental toolkit for translocon pore analysis to include state-of-the-art in vitro biochemical experiments based on various nano-membrane environments, including nanolipoprotein particles [86] and black lipid bilayers [70], which could be optimized for compatibility with new-age imaging technologies. Significantly, powerful imaging technologies are already being applied with impressive results to study pore-induced alterations in morphology and deformability of RBCs [87]. Conceivably, pore-forming proteins could even be utilized to stimulate aged-related mechanisms of RBC turnover that would help to further probe the molecular processes of membrane shedding and eryptosis in these cells.

## 2 Materials

### 2.1 Red Blood Cells

Freshly purchased sheep red blood cells (RBCs) (*see* **Note 1**) in Alsever's buffer (*see* **Note 2**).

### 2.2 Strains

The *Y. pseudotuberculosis* Δ*yopK* null mutant (YPIII/pIB155) is routinely used as the parent strain into which other mutations of interest are introduced. We routinely use the pDM4-based method to assist with the *in cis* generation of mutants by allelic exchange [88]. The Δ*yopK* null mutant was shown to augment RBC lysis [44]. The precise molecular basis for this effect is not fully understood, but translocated YopK is involved in a feedback inhibitory loop to limit excessive pore formation [43–46]. Using this genetic background is therefore a convenient approach to enhance the sensitivity of the contact-dependent erythrocyte hemolysis readout (*see* **Note 3**).

### 2.3 Growth Media

*Y. pseudotuberculosis* bacteria are cultured in Brain Heart Infusion (BHI) broth supplemented with 5 mM of EGTA and 20 mM

$MgCl_2$ and in the presence of appropriate antibiotics. Critically, this generates a liquid growth medium that is devoid of calcium ions (BHI–$Ca^{2+}$). When culturing in this media at an incubation temperature of 37 °C, the synthesis and secretion of Yops by *Yersinia* bacteria is specifically induced. An alternative media can be based on a version of Lysogeny broth (LB) (*see* **Note 4**) that is similarly depleted of calcium ions by supplementation with 5 mM of EGTA and 20 mM $MgCl_2$.

#### 2.4 Specific Reagents for Osmoprotection Assays

To ensure that hemolysis is caused by pores of similar dimension as a result of the ordered trans-membrane insertion of YopB and YopD, we often perform in parallel a second contact-dependent hemolysis assay in the presence of oligosaccharides of different sizes (Fig. 1d) (*see* **Note 5**). Routinely, we use three different-sized sugars. The smallest-sized sugar is Raffinose (1.2–1.4 nm ϕ), which cannot effectively block hemolytic activity induced by a standard sized translocon pore. Use of two larger sugars—Dextrin 15 (2.2 nm ϕ) and Dextran 4 (3–3.5 nm ϕ)—progressively inhibits more of the hemolytic activity induced by a standard-sized translocon pore.

#### 2.5 Consumables and Equipment for Lysis Assays and Osmoprotection Assays

- Sterile 15 mL cell culture tubes.
- 96-well round-bottom microplates.
- 96-well flat-bottom assay microplates compatible with automated readers.
- Microplate reader spectrophotometer.
- Benchtop centrifuge with capacity for culture tube and microtiter plate centrifugation.

#### 2.6 Membrane-Enrichment Assay Reagents

- 5 mM Tris–HCl (pH 7.4).
- 10× TBS (200 mM Tris–HCl (pH 7.4), 1.5 M NaCl).
- Protease inhibitor cocktail.
- 25% Sucrose-TBS (20 mM Tris–HCl (pH 7.4), 150 mM NaCl).
- 44% Sucrose-TBS (20 mM Tris–HCl (pH 7.4), 150 mM NaCl).
- 50% Sucrose-TBS (20 mM Tris–HCl (pH 7.4), 150 mM NaCl).
- 65% Sucrose-TBS (20 mM Tris–HCl (pH 7.4), 150 mM NaCl).
- SDS-PAGE and immunoblotting reagents (including antigen-specific antibodies).

#### 2.7 Consumables and Equipment for Membrane-Enrichment Assays

- Sterile 50 mL cell culture tubes.
- Sonicator.
- Polycarbonate ultracentrifuge tubes.

- Ultracentrifuge and accompanying suitable fixed-angle and swing-out rotors.
- SDS-PAGE and immunoblotting equipment (including Immobilon-P PVDF membrane).

**2.8 Membrane Protein Analysis Reagents**

- 100 mM $NaCO_3$ (pH 10).
- 20% Dodecyl-β-D-maltoside.
- 8 M Urea.
- Coomassie blue stain.

## 3 Methods

**3.1 Contact-Dependent Erythrocyte Lysis**

1. The *Y. pseudotuberculosis* mutant bacteria are grown with agitation overnight at 26 °C in 5 mL BHI–$Ca^{2+}$ broth with appropriated presence of antibiotics.
2. Overnight bacterial cultures are diluted 10× in fresh BHI–$Ca^{2+}$ medium and grown with agitation for 1 h at 26 °C followed by 1 h incubation at 37 °C.
3. Bacterial cultures are standardized with culture media to the same absorbance (between 0.7 and 0.9) by spectrophotometric measurement at 600 nm (*see* **Note 6**).
4. Sheep erythrocytes are washed three times with cold PBS and then adjusted to $4 \times 10^9$ cells/mL with BHI–$Ca^{2+}$ medium preheated at 37 °C (*see* **Note 7**).
5. In quadruplicate per strain (i.e., four technical replicates), 50 μL of bacterial culture and 50 μL of RBCs suspension are mixed in a 96-well round-bottom microplate and centrifuged at 1200 × *g* for 5 min at room temperature (*see* **Notes 8–10**).
6. Incubate the microplate for 90 min at 37 °C in normal atmospheric conditions, and then carefully resuspend each of the pellets with 150 μL of cold PBS (*see* **Note 11**).
7. Centrifuge the microplate at 1200 × *g* for 15 min at 4 °C and carefully transfer 150 μL of the supernatant from each well into a 96-well flat-bottom microplate.
8. The amount of hemoglobin released is obtained by spectrophotometric measurement at a wavelength of 545 nm.

**3.2 Contact-Dependent Erythrocyte Lysis in the Presence of Carbohydrates: An Osmoprotection Assay**

1. Follow all steps in Subheading 3.1 above, but this time add the respective oligosaccharides (e.g., Raffinose, Dextrin 15, -Dextran 4) at a final concentration of 30 mM in the solutions used for washing the RBCs (as performed in Subheading 3.1, **step 4** above) and for resuspending the infected RBC pellet (as performed in Subheading 3.1, **step 6** above) (*see* **Note 5**).

### 3.3 Purification of Sheep Erythrocyte Membranes

1. Grow overnight *Y. pseudotuberculosis* bacteria with agitation at 26 °C in 5 mL fresh LB160 liquid media (*see* **Note 4**) that is also supplemented with 20 mM $MgCl_2$ and 5 mM EGTA to achieve a calcium-depleted media that we term LB160–$Ca^{2+}$. If appropriate, antibiotics can also be added to this media.
2. In 50 mL conical tubes, subculture 0.1 volumes of overnight grown bacterial culture into 10 mL fresh LB160–$Ca^{2+}$ media, but that now additionally contained a protease-inhibitor cocktail (*see* **Note 12**).
3. Grow the cultures with agitation for 1 h at 26 °C and then for 1 h at 37 °C, then standardize with fresh LB160–$Ca^{2+}$ media to the same absorbance (between 0.7 and 0.9) by spectrophotometric measurement at 600 nm.
4. Using a benchtop centrifuge at 3000 × *g* for 5 min, harvest the bacteria into a pellet that would correspond to a total of $5 \times 10^9$ to $2 \times 10^{10}$ bacteria.
5. Resuspend the bacterial pellet in 5 mL washed erythrocytes (*see* Subheading 3.1, **step 4** and **Note 7** for washing instructions) corresponding to $2 \times 10^{11}$ cells contained in the LB160–$Ca^{2+}$ media supplemented with protease-inhibitor cocktail (*see* **Note 12**).
6. Centrifuge the mixture at 3500 × *g* for 5 min to create contact between bacteria and cells, and then incubate for 2 h at 37 °C in normal atmospheric conditions (*see* **Notes 10** and **11**).
7. To isolate the erythrocytes membranes following hemolysis (*see* **Note 13**), add 35 mL of ice-cold 5 mM Tris–HCl (pH 7.4) (with protease-inhibitor cocktail) to the mix. Mix thoroughly and then incubate on ice for 10 min.
8. Supplement each reaction with 4.5 mL of 10× TBS (*see* **Note 14**), and then centrifuge at 3500 × *g* for 20 min at 4 °C.
9. To harvest the membranes, transfer the supernatants in a volume of 42 mL to 38 × 102 mm open-top, thick-wall polycarbonate tubes (Beckman) and centrifuge at 100,000 × *g* for 2 h at 4 °C.
10. Resuspend the membrane pellets in 10 mL of 50% sucrose-TBS solution containing protease-inhibitor cocktail by stirring and gentle sonication in an ultrasonic bath.
11. Establish a step-wise gradient in ultra-clear 25 × 89 mm centrifugation tubes by overlaying resuspended samples on 2 mL 65% sucrose in TBS and then onto which overlay with 15 mL of 44% sucrose in TBS followed by 5 mL of 25% sucrose in TBS (*see* **Note 15**).
12. Centrifuged at 15,000 × *g* for 16 h at 4 °C in a swing-out rotor (*see* **Note 16**).

13. Recover the enriched erythrocyte membrane fraction from the 25–44% sucrose-TBS interface into five volumes of TBS and centrifuge at 100,000 × *g* for 2 h at 4 °C.
14. Resuspend the washed and concentrated membranes in 150 μL of TBS.
15. Fractionate on a 12% SDS-PAGE the equivalent of 20 μg of protein from each sample (*see* **Note 17**) and using a standard Western blotting procedure with Immobilon-P PVDF membrane, analyze for the presence of YopB and YopD using specific anti-YopB and anti-YopD polyclonal antibodies raised in rabbits (*see* **Note 18**).

### 3.4 Separation of Integrated and Associated Membrane Proteins

1. To analyze the nature of YopB and YopD association with the erythrocyte membrane, dilute 100 μg of purified membranes 100-fold with 100 mM $NaCO_3$ (pH 10) and incubate on ice for 1 h with gentle stirring. Resistance to sodium carbonate extraction is an established method to distinguish integral membrane proteins from membrane-associated proteins (*see* **Note 19**).
2. Recover the membrane fractions by ultracentrifugation at 100,000 × *g* for 2 h.
3. To analyze the isolated membranes in the pelleted fraction and the soluble proteins recovered in the supernatants, load the samples on two duplicate 12% acrylamide SDS-PAGE gels. Transfer one of the gels to an Immobilon-P PVDF membrane (Millipore) and analyze by Western blot with anti-YopB and anti-YopD antibodies. Subject the second gel to Coomassie-staining to serve as a loading control (*see* **Note 20**).

## 4 Notes

1. We routinely use sheep as the animal source of the erythrocytes. Erythrocytes from other animal species could have an impact on the assay readout. The molecular basis for why *Yersinia* more efficiently lyses and/or yields more released hemoglobin upon infections of sheep erythrocytes compared to other animal or human origin is not understood. Actually, it is generally accepted that the species of erythrocyte, or even erythrocytes from different individuals of the same species can vary in their sensitivity to a pore-forming toxin [62]. In fact, there is very little understanding concerning comparative physiological differences between populations of RBCs originating from within the same individual, different individuals of the same species, or individuals from different species [89]. As an aside, it is anticipated that unlocking the secrets of this heterogeneity among

and between populations of RBCs can provide a better understanding of health and disease.

2. To ensure a clear readout during the contact-dependent hemolysis assay, always use fresh RBCs preserved in commercially available Alsever's buffer (2.05% dextrose, 0.8% sodium citrate, 0.055% citric acid, and 0.42% sodium chloride). Alsever's solution functions as a blood preservative and anticoagulant.

3. In addition to the augmentation of contact-dependent RBC lysis by genetic means through utilizing a *yopK* mutant, chemical augmentation is possible via exposure of the hemolytic reaction to a mild hypotonic solution (final concentration 0.75× PBS). This yields an osmotic pressure of 235.05 mOsm/L, which compares to 313.4 mOsm/L for standard "isotonic" PBS [0.02% (w/v) Potassium chloride, 0.02% (w/v) Potassium dihydrogen phosphate, 0.8% (w/v) Sodium chloride, 0.115% (w/v) Disodium hydrogen phosphate (anhydrous)]. Importantly, this modest increase in osmotic pressure does not induce unspecific erythrocyte lysis. (*See* also **Note 11** for another effective augmentation strategy.)

4. As an alternative bacterial culture medium, it is appropriate to use calcium ion-depleted Lysogeny broth (LB) that is achieved through supplementation with 5 mM of EGTA and 20 mM $MgCl_2$. Note that the original LB media contains 160 mM NaCl [90], while other variants also exist that contain NaCl at lower concentrations. Hence, to indicate that we deliberately employ LB media with 160 mM NaCl, we use the designation LB160 throughout this chapter (*see* Subheading 3.3). It follows that if required, we recommend to supplement available LB media with the necessary amount of NaCl to achieve a final concentration of 160 mM.

5. Oligosaccharides of increasing size are added to the assay at 30 mM concentration as this corresponds to the internal osmotic pressure of the RBCs. The prevention of hemolysis by oligosaccharides is indicative of pore formation of a restricted size, as distinct from hemolysis caused by other forms of membrane damage [91]. If the oligosaccharide is too small to block the pore, it will diffuse through the pore along with water, and this will cause swelling, osmolysis, and hemoglobin release. In contrast, a suitably large oligosaccharide that is bulky enough to block the entire pore will prevent water influx into the cell, counterbalancing the internal pressure, such that the infected RBC will be osmoprotected and hemoglobin release inhibited.

6. Optimal hemolysis is routinely achieved by utilizing bacteria grown to mid exponential growth phase.

7. Gentle washing is critical to maintain erythrocyte cell integrity. The washed RBCs supernatant must be clear from any visible hemoglobin contamination. Continue with gentle washing until this has been achieved. If this point is not reached following two to four gentle washes, it is a clear sign that the RBC population is too old. We routinely adjust the washed RBC population density to a concentration of $4 \times 10^9$ cells/mL with 5 mL of fresh BHI–$Ca^{2+}$ medium.
8. A blank control where only BHI–$Ca^{2+}$ or LB160–$Ca^{2+}$ is added to the RBCs must be incorporated into each biological replicate to estimate the extent of spontaneous background lysis. It is recommended that a minimum of five independent biological replicates are performed, with each biological replicate being measured in four technical replicates.
9. While the classical protocol dictates a 0.1 mL final volume of bacteria and RBC mix incubated in a microtiter tray, the protocol is amenable to a tenfold scale-up whereby a 1 mL mix of bacteria and RBCs is incubated in single-use 15 mL plastic tubes. In this case, the vicinity of $2 \times 10^{11}$ cells/mL are used for each single reaction.
10. Centrifugation is needed to bring the bacteria into close contact with the red blood cell. For those bacteria with a short protruding T3SS needle—that is typical of the human pathogenic *Yersinia* sp., omitting this centrifugation step will negatively affect the efficiency of hemolysis [35, 64, 92].
11. Perform intermittent resuspension and centrifugation every 30 min to increase the contact-dependent hemolysis. Presumably, breaking the bacterial contact with the RBC leaves exposed pores in the membrane that permits water influx resulting in enhanced osmotic lysis and hemoglobin release. (*See* also **Note 3** for another augmentation strategy.)
12. We routinely use cOmplete™ Mini Protease Inhibitor Cocktail EDTA-free (Roche). From this point, this protease-inhibitor cocktail was included in all solutions.
13. Hemolysis can be confirmed using the following analysis. After 2 h, sample 100 μL volumes from the reaction mixes, mix with an equal volume of PBS, and then centrifuge at $1500 \times g$ for 5 min. Transfer 100 μL of the supernatant to a flat bottom plate and measure the absorbance at 545 nm to determine the extent of hemolysis.
14. We deliberately added 10× TBS solution to restore the pH and salt concentration in the samples.
15. The procedure to enrich for erythrocyte membrane from hemolytic reactions by floatation density gradient purification is adapted from earlier studies [45, 80].

16. As an example, we use the SW-28 rotor from Beckman-Coulter.
17. Protein concentrations were estimated by measuring $A_{280}$ on a NanoDrop spectrophotometer.
18. It is critical to test samples from different steps in the purification protocol for the presence of contaminating bacterial proteins. As an example, we have utilized rabbit polyclonal antibodies specific for the bacterial membrane protein OmpF and specific for the soluble cytoplasmic protein DnaJ to confirm the absence of contaminating bacterial proteins in the purified erythrocyte membrane fraction after infection with parent *Y. pseudotuberculosis*. To test the samples for the presence of contaminating soluble eukaryotic protein, we have taken advantage of the fact that our anti-DnaJ antibodies also recognize an erythrocyte-derived soluble protein with a size around 40 kDa that is likely to be Hsp40, the eukaryotic homologue of DnaJ [93]. Hence, the absence of Hsp40 in the final membrane preparations would indicate that the samples are clear of contaminating soluble eukaryotic protein. On the other hand, to confirm the purification of sheep erythrocyte membrane material, we assay for the enrichment of the erythrocyte-specific integral membrane protein $Na^+/K^+$ ATPase by detection with an anti-$Na^+/K^+$ ATPase α-1 monoclonal antibody (clone C464.6; Millipore).
19. As a follow-up to determine if YopB and YopD are truly integrated in the purified erythrocyte membranes, it is also recommended to solubilize 100 μg of purified membranes with a final concentration of 2% (w/v) Dodecyl-β-D-maltoside for 1 h with gentle stirring on ice. Solubilization of biological membranes with a suitable non-ionic detergent solubilizes integral membrane proteins from the membrane vesicles. Moreover, to exclude that YopB and YopD accumulate in the erythrocyte membrane fraction by aggregation or precipitation during the purification procedure, it is also recommended that 100 μg of purified membranes be extracted with a final concentration of 4 M Urea for 1 h with gentle stirring on ice. Washing the membranes with Urea will release peripheral membrane proteins and any aggregates that have accumulated during the membrane purification procedure, while integral membrane proteins will remain in the hydrophobic membrane fraction. As already described, the membrane fractions are recovered by ultracentrifugation at 100,000 × *g* for 2 h, and the material subjected to SDS-PAGE and western blotting.
20. A couple of independent studies have now shown that native YopD and YopB are integrally inserted into eythrocyte membranes as they resist thorough membrane washing with either

sodium carbonate or urea, while they were solubilized with Dodecyl-β-D-maltoside detergent [34, 69]. Interestingly, we have found that the majority of YopD solubilize from membrane after urea washes following infection with a Δ*yopB* null mutant, but this same treatment still leaves YopB membrane integrated following infection with a Δ*yopD* null mutant [69]. Hence, proper YopD membrane integration is YopB dependent, but YopB membrane integration is not dependent on YopD to the same extent. Interestingly, this corroborates a recent study with *Pseudomonas aeruginosa* where the YopB homologue—PopB—was found to assist with the proper insertion of the YopD homologue—PopD—into plasma membranes of nucleated cells [94]. However, we also believe that there must be some cooperativity in translocon assembly since the full-length mutants of Δ*yopB* and Δ*yopD* could only accommodate the insertion of modest levels of the remaining translocator [69]. Expectantly, the levels of membrane-integrated YopB and YopD increase dramatically when using an isogenic mutant lacking *yopK* [69], coding for an established regulator of pore formation and effector translocation [43–46].

## Acknowledgements

This work was performed within the framework of the Umeå Centre for Microbial Research at Umeå University. MSF acknowledges financial support from the Swedish Research Council grant 2014–2105, the Medical Research Foundation of Umeå University, and the Faculty of Science and Technology at Umeå University.

### References

1. Buttner D (2012) Protein export according to schedule: architecture, assembly, and regulation of type III secretion systems from plant- and animal-pathogenic bacteria. Microbiol Mol Biol Rev 76(2):262–310
2. Deng W, Marshall NC, Rowland JL, McCoy JM, Worrall LJ, Santos AS, Strynadka NCJ, Finlay BB (2017) Assembly, structure, function and regulation of type III secretion systems. Nat Rev Microbiol 15(6):323–337. https://doi.org/10.1038/nrmicro.2017.20
3. Pallen MJ, Beatson SA, Bailey CM (2005) Bioinformatics, genomics and evolution of non-flagellar type-III secretion systems: a Darwinian perpective. FEMS Microbiol Rev 29 (2):201–229
4. Erhardt M, Namba K, Hughes KT (2010) Bacterial nanomachines: the flagellum and type III injectisome. Cold Spring Harb Perspect Biol 2 (11):a000299
5. Diepold A, Armitage JP (2015) Type III secretion systems: the bacterial flagellum and the injectisome. Philos Trans R Soc Lond B Biol Sci 370(1679):pii: 20150020. https://doi.org/10.1098/rstb.2015.0020
6. Hu B, Lara-Tejero M, Kong Q, Galan JE, Liu J (2017) *In situ* molecular architecture of the *Salmonella* type III secretion machine. Cell 168(6):1065–1074.e1010. https://doi.org/10.1016/j.cell.2017.02.022
7. Worrall LJ, Hong C, Vuckovic M, Deng W, Bergeron JR, Majewski DD, Huang RK, Spreter T, Finlay BB, Yu Z, Strynadka NC (2016) Near-atomic-resolution cryo-EM analysis of the *Salmonella* T3S injectisome basal body. Nature 540:597. https://doi.org/10.1038/nature20576

8. Nans A, Kudryashev M, Saibil HR, Hayward RD (2015) Structure of a bacterial type III secretion system in contact with a host membrane *in situ*. Nat Commun 6:10114. https://doi.org/10.1038/ncomms10114
9. Radics J, Konigsmaier L, Marlovits TC (2014) Structure of a pathogenic type 3 secretion system in action. Nat Struct Mol Biol 21 (1):82–87. https://doi.org/10.1038/nsmb.2722
10. Dohlich K, Zumsteg AB, Goosmann C, Kolbe M (2014) A substrate-fusion protein is trapped inside the type III secretion system channel in *Shigella flexneri*. PLoS Pathog 10(1): e1003881. https://doi.org/10.1371/journal.ppat.1003881
11. Akeda Y, Galan JE (2005) Chaperone release and unfolding of substrates in type III secretion. Nature 437(7060):911–915
12. Lee PC, Rietsch A (2015) Fueling type III secretion. Trends Microbiol 23(5):296–300. https://doi.org/10.1016/j.tim.2015.01.012
13. Erhardt M, Mertens ME, Fabiani FD, Hughes KT (2014) ATPase-independent type-III protein secretion in *Salmonella enterica*. PLoS Genet 10(11):e1004800. https://doi.org/10.1371/journal.pgen.1004800
14. Wilharm G, Dittmann S, Schmid A, Heesemann J (2007) On the role of specific chaperones, the specific ATPase, and the proton motive force in type III secretion. Int J Med Microbiol 297(1):27–36
15. Mahdavi A, Szychowski J, Ngo JT, Sweredoski MJ, Graham RL, Hess S, Schneewind O, Mazmanian SK, Tirrell DA (2014) Identification of secreted bacterial proteins by noncanonical amino acid tagging. Proc Natl Acad Sci U S A 111(1):433–438. https://doi.org/10.1073/pnas.1301740111
16. Barison N, Gupta R, Kolbe M (2013) A sophisticated multi-step secretion mechanism: how the type 3 secretion system is regulated. Cell Microbiol 15(11):1809–1817. https://doi.org/10.1111/cmi.12178
17. Dewoody RS, Merritt PM, Marketon MM (2013) Regulation of the *Yersinia* type III secretion system: traffic control. Front Cell Infect Microbiol 3:4
18. Galan JE (2009) Common themes in the design and function of bacterial effectors. Cell Host Microbe 5(6):571–579
19. Dean P (2011) Functional domains and motifs of bacterial type III effector proteins and their roles in infection. FEMS Microbiol Rev 35 (6):1100–1125
20. Grabowski B, Schmidt MA, Ruter C (2017) Immunomodulatory *Yersinia* outer proteins (Yops)-useful tools for bacteria and humans alike. Virulence 8(7):1124–1147. https://doi.org/10.1080/21505594.2017.1303588
21. Mattei PJ, Faudry E, Job V, Izore T, Attree I, Dessen A (2011) Membrane targeting and pore formation by the type III secretion system translocon. FEBS J 278(3):414–426
22. Mueller CA, Broz P, Cornelis GR (2008) The type III secretion system tip complex and translocon. Mol Microbiol 68(5):1085–1095
23. Tejeda-Dominguez F, Huerta-Cantillo J, Chavez-Duenas L, Navarro-Garcia F (2017) A novel mechanism for protein delivery by the type 3 secretion system for extracellularly secreted proteins. MBio 8(2):pii: e00184-17. https://doi.org/10.1128/mBio.00184-17
24. Akopyan K, Edgren T, Wang-Edgren H, Rosqvist R, Fahlgren A, Wolf-Watz H, Fallman M (2011) Translocation of surface-localized effectors in type III secretion. Proc Natl Acad Sci U S A 108(4):1639–1644
25. Russo BC, Stamm LM, Raaben M, Kim CM, Kahoud E, Robinson LR, Bose S, Queiroz AL, Herrera BB, Baxt LA, Mor-Vaknin N, Fu Y, Molina G, Markovitz DM, Whelan SP, Goldberg MB (2016) Intermediate filaments enable pathogen docking to trigger type 3 effector translocation. Nat Microbiol 1:16025. https://doi.org/10.1038/nmicrobiol.2016.25
26. Armentrout EI, Rietsch A (2016) The type III secretion translocation pore senses host cell contact. PLoS Pathog 12(3):e1005530. https://doi.org/10.1371/journal.ppat.1005530
27. Ji H, Dong H (2015) Key steps in type III secretion system (T3SS) towards translocon assembly with potential sensor at plant plasma membrane. Mol Plant Pathol 16(7):762–773. https://doi.org/10.1111/mpp.12223
28. Sheahan KL, Isberg RR (2015) Identification of mammalian proteins that collaborate with type III secretion system function: involvement of a chemokine receptor in supporting translocon activity. MBio 6(1):e02023-02014. https://doi.org/10.1128/mBio.02023-14
29. Viala JP, Prima V, Puppo R, Agrebi R, Canestrari MJ, Lignon S, Chauvin N, Meresse S, Mignot T, Lebrun R, Bouveret E (2017) Acylation of the type 3 secretion system translocon using a dedicated acyl carrier protein. PLoS Genet 13(1):e1006556. https://doi.org/10.1371/journal.pgen.1006556
30. Dortet L, Lombardi C, Cretin F, Dessen A, Filloux A (2018) Pore-forming activity of the *Pseudomonas aeruginosa* type III secretion system translocon alters the host epigenome. Nat

Microbiol 3(3):378–386. https://doi.org/10.1038/s41564-018-0109-7

31. Francis MS, Wolf-Watz H (1998) YopD of *Yersinia pseudotuberculosis* is translocated into the cytosol of HeLa epithelial cells: evidence of a structural domain necessary for translocation. Mol Microbiol 29(3):799–813
32. Neyt C, Cornelis GR (1999) Insertion of a Yop translocation pore into the macrophage plasma membrane by *Yersinia enterocolitica*: requirement for translocators YopB and YopD, but not LcrG. Mol Microbiol 33(5):971–981
33. Marenne MN, Journet L, Mota LJ, Cornelis GR (2003) Genetic analysis of the formation of the Ysc-Yop translocation pore in macrophages by *Yersinia enterocolitica*: role of LcrV, YscF and YopN. Microb Pathog 35(6):243–258
34. Montagner C, Arquint C, Cornelis GR (2011) Translocators YopB and YopD from *Yersinia* form a multimeric integral membrane complex in eukaryotic cell membranes. J Bacteriol 193 (24):6923–6928
35. Håkansson S, Schesser K, Persson C, Galyov EE, Rosqvist R, Homble F, Wolf-Watz H (1996) The YopB protein of *Yersinia pseudotuberculosis* is essential for the translocation of Yop effector proteins across the target cell plasma membrane and displays a contact-dependent membrane disrupting activity. EMBO J 15(21):5812–5823
36. Tardy F, Homble F, Neyt C, Wattiez R, Cornelis GR, Ruysschaert JM, Cabiaux V (1999) *Yersinia enterocolitica* type III secretion-translocation system: channel formation by secreted Yops. EMBO J 18(23):6793–6799
37. Håkansson S, Bergman T, Vanooteghem JC, Cornelis G, Wolf-Watz H (1993) YopB and YopD constitute a novel class of *Yersinia* Yop proteins. Infect Immun 61(1):71–80
38. Mueller CA, Broz P, Muller SA, Ringler P, Erne-Brand F, Sorg I, Kuhn M, Engel A, Cornelis GR (2005) The V-antigen of *Yersinia* forms a distinct structure at the tip of injectisome needles. Science 310(5748):674–676
39. Broz P, Mueller CA, Muller SA, Philippsen A, Sorg I, Engel A, Cornelis GR (2007) Function and molecular architecture of the *Yersinia* injectisome tip complex. Mol Microbiol 65 (5):1311–1320
40. Goure J, Broz P, Attree O, Cornelis GR, Attree I (2005) Protective anti-V antibodies inhibit *Pseudomonas* and *Yersinia* translocon assembly within host membranes. J Infect Dis 192 (2):218–225
41. Ligtenberg KG, Miller NC, Mitchell A, Plano GV, Schneewind O (2013) LcrV mutants that abolish *Yersinia* type III injectisome function. J Bacteriol 195(4):777–787. https://doi.org/10.1128/JB.02021-12
42. Ekestubbe S, Broms JE, Edgren T, Fallman M, Francis MS, Forsberg A (2016) The amino-terminal part of the needle-tip translocator LcrV of *Yersinia pseudotuberculosis* is required for early targeting of YopH and *in vivo* virulence. Front Cell Infect Microbiol 6:175. https://doi.org/10.3389/fcimb.2016.00175
43. Dewoody R, Merritt PM, Marketon MM (2013) YopK controls both rate and fidelity of Yop translocation. Mol Microbiol 87 (2):301–317
44. Holmström A, Pettersson J, Rosqvist R, Håkansson S, Tafazoli F, Fällman M, Magnusson KE, Wolf-Watz H, Forsberg Å (1997) YopK of *Yersinia pseudotuberculosis* controls translocation of Yop effectors across the eukaryotic cell membrane. Mol Microbiol 24 (1):73–91
45. Thorslund SE, Edgren T, Pettersson J, Nordfelth R, Sellin ME, Ivanova E, Francis MS, Isaksson EL, Wolf-Watz H, Fallman M (2011) The RACK1 signaling scaffold protein selectively interacts with *Yersinia pseudotuberculosis* virulence function. PLoS One 6(2): e16784
46. Dewoody R, Merritt PM, Houppert AS, Marketon MM (2011) YopK regulates the *Yersinia pestis* type III secretion system from within host cells. Mol Microbiol 79(6):1445–1461
47. Costa TR, Edqvist PJ, Broms JE, Ahlund MK, Forsberg A, Francis MS (2010) YopD self-assembly and binding to LcrV facilitate type III secretion activity by *Yersinia pseudotuberculosis*. J Biol Chem 285(33):25269–25284
48. Brodsky IE, Palm NW, Sadanand S, Ryndak MB, Sutterwala FS, Flavell RA, Bliska JB, Medzhitov R (2010) A *Yersinia* effector protein promotes virulence by preventing inflammasome recognition of the type III secretion system. Cell Host Microbe 7(5):376–387
49. Boyd AP, Grosdent N, Totemeyer S, Geuijen C, Bleves S, Iriarte M, Lambermont I, Octave JN, Cornelis GR (2000) *Yersinia enterocolitica* can deliver Yop proteins into a wide range of cell types: development of a delivery system for heterologous proteins. Eur J Cell Biol 79(10):659–671
50. Durand EA, Maldonado-Arocho FJ, Castillo C, Walsh RL, Mecsas J (2010) The presence of professional phagocytes dictates the number of host cells targeted for Yop translocation during infection. Cell Microbiol 12 (8):1064–1082
51. Koberle M, Klein-Gunther A, Schutz M, Fritz M, Berchtold S, Tolosa E, Autenrieth

IB, Bohn E (2009) *Yersinia enterocolitica* targets cells of the innate and adaptive immune system by injection of Yops in a mouse infection model. PLoS Pathog 5(8):e1000551
52. Marketon MM, DePaolo RW, DeBord KL, Jabri B, Schneewind O (2005) Plague bacteria target immune cells during infection. Science 309(5741):1739–1741
53. Ashton N (2013) Physiology of red and white blood cells. Anaesth Intens Care Med 14 (6):261–266. https://doi.org/10.1016/j.mpaic.2013.03.001
54. Gordon-Smith T (2013) Structure and function of red and white blood cells. Medicine 41 (4):193–199. https://doi.org/10.1016/j.mpmed.2013.01.023
55. Bain BJ (2017) Structure and function of red and white blood cells. Medicine 45 (4):187–193. https://doi.org/10.1016/j.mpmed.2017.01.011
56. Diez-Silva M, Dao M, Han J, Lim CT, Suresh S (2010) Shape and biomechanical characteristics of human red blood cells in health and disease. MRS Bull 35(5):382–388
57. Viallat A, Abkarian M (2014) Red blood cell: from its mechanics to its motion in shear flow. Int J Lab Hematol 36(3):237–243. https://doi.org/10.1111/ijlh.12233
58. SEt L (2016) Anatomy of the red cell membrane skeleton: unanswered questions. Blood 127(2):187–199. https://doi.org/10.1182/blood-2014-12-512772
59. Kaestner L, Minetti G (2017) The potential of erythrocytes as cellular aging models. Cell Death Differ 24(9):1475–1477. https://doi.org/10.1038/cdd.2017.100
60. Kay MM, Goodman J (2003) Immunoregulation of cellular lifespan: physiologic autoantibodies and their peptide antigens. Cell Mol Biol (Noisy-le-Grand) 49(2):217–243
61. Pretorius E, du Plooy JN, Bester J (2016) A comprehensive review on eryptosis. Cell Physiol Biochem 39(5):1977–2000. https://doi.org/10.1159/000447895
62. Bernheimer AW (1988) Assay of hemolytic toxins. Methods Enzymol 165:213–217. https://doi.org/10.1016/S0076-6879(88)65033-6
63. Rowe GE, Welch RA (1994) Assays of hemolytic toxins. Methods Enzymol 235:657–667
64. Clerc P, Baudry B (1985) Sansonetti PJ (1986) Plasmid-mediated contact haemolytic activity in *Shigella* species: correlation with penetration into HeLa cells. Ann Inst Pasteur Microbiol 137A(3):267–278
65. Dacheux D, Goure J, Chabert J, Usson Y, Attree I (2001) Pore-forming activity of type III system-secreted proteins leads to oncosis of *Pseudomonas aeruginosa*-infected macrophages. Mol Microbiol 40(1):76–85
66. Warawa J, Finlay BB, Kenny B (1999) Type III secretion-dependent hemolytic activity of enteropathogenic *Escherichia coli*. Infect Immun 67(10):5538–5540
67. Kwuan L, Adams W, Auerbuch V (2013) Impact of host membrane pore formation by the *Yersinia pseudotuberculosis* type III secretion system on the macrophage innate immune response. Infect Immun 81(3):905–914. https://doi.org/10.1128/IAI.01014-12
68. Olsson J, Edqvist PJ, Bröms JE, Forsberg Å, Wolf-Watz H, Francis MS (2004) The YopD translocator of *Yersinia pseudotuberculosis* is a multifunctional protein comprised of discrete domains. J Bacteriol 186(13):4110–4123
69. Costa TR, Amer AA, Farag SI, Wolf-Watz H, Fallman M, Fahlgren A, Edgren T, Francis MS (2013) Type III secretion translocon assemblies that attenuate *Yersinia* virulence. Cell Microbiol 15(7):1088–1110
70. Holmström A, Olsson J, Cherepanov P, Maier E, Nordfelth R, Pettersson J, Benz R, Wolf-Watz H, Forsberg Å (2001) LcrV is a channel size-determining component of the Yop effector translocon of *Yersinia*. Mol Microbiol 39(3):620–632
71. Bröms JE, Sundin C, Francis MS, Forsberg Å (2003) Comparative analysis of type III effector translocation by *Yersinia pseudotuberculosis* expressing native LcrV or PcrV from *Pseudomonas aeruginosa*. J Infect Dis 188(2):239–249
72. Zwack EE, Snyder AG, Wynosky-Dolfi MA, Ruthel G, Philip NH, Marketon MM, Francis MS, Bliska JB, Brodsky IE (2015) Inflammasome activation in response to the *Yersinia* type III secretion system requires hyperinjection of translocon proteins YopB and YopD. MBio 6 (1):e02095-02014. https://doi.org/10.1128/mBio.02095-14
73. Edqvist PJ, Aili M, Liu J, Francis MS (2007) Minimal YopB and YopD translocator secretion by *Yersinia* is sufficient for Yop-effector delivery into target cells. Microbes Infect 9 (2):224–233
74. Auerbuch V, Golenbock DT, Isberg RR (2009) Innate immune recognition of *Yersinia pseudotuberculosis* type III secretion. PLoS Pathog 5 (12):e1000686. https://doi.org/10.1371/journal.ppat.1000686
75. Viboud GI, So SS, Ryndak MB, Bliska JB (2003) Proinflammatory signalling stimulated by the type III translocation factor YopB is counteracted by multiple effectors in epithelial

cells infected with *Yersinia pseudotuberculosis*. Mol Microbiol 47(5):1305–1315
76. Viboud GI, Bliska JB (2001) A bacterial type III secretion system inhibits actin polymerization to prevent pore formation in host cell membranes. EMBO J 20(19):5373–5382
77. Shaw RK, Daniell S, Frankel G, Knutton S (2002) Enteropathogenic *Escherichia coli* translocate Tir and form an intimin-Tir intimate attachment to red blood cell membranes. Microbiology 148(Pt 5):1355–1365
78. Ciana A, Achilli C, Minetti G (2017) Spectrin and other membrane-skeletal components in human red blood cells of different age. Cell Physiol Biochem 42(3):1139–1152. https://doi.org/10.1159/000478769
79. Romero M, Keyel M, Shi G, Bhattacharjee P, Roth R, Heuser JE, Keyel PA (2017) Intrinsic repair protects cells from pore-forming toxins by microvesicle shedding. Cell Death Differ 24 (5):798–808. https://doi.org/10.1038/cdd.2017.11
80. Blocker A, Gounon P, Larquet E, Niebuhr K, Cabiaux V, Parsot C, Sansonetti P (1999) The tripartite type III secreton of *Shigella flexneri* inserts IpaB and IpaC into host membranes. J Cell Biol 147(3):683–693
81. Mejia E, Bliska JB, Viboud GI (2008) *Yersinia* controls type III effector delivery into host cells by modulating Rho activity. PLoS Pathog 4(1): e3. https://doi.org/10.1371/journal.ppat.0040003
82. Solomon R, Zhang W, McCrann G, Bliska JB, Viboud GI (2015) Random mutagenesis identifies a C-terminal region of YopD important for *Yersinia* type III secretion function. PLoS One 10(3):e0120471. https://doi.org/10.1371/journal.pone.0120471
83. Shin H, Cornelis GR (2007) Type III secretion translocation pores of *Yersinia enterocolitica* trigger maturation and release of pro-inflammatory IL-1beta. Cell Microbiol 9 (12):2893–2902. https://doi.org/10.1111/j.1462-5822.2007.01004.x
84. Adams W, Morgan J, Kwuan L, Auerbuch V (2015) *Yersinia pseudotuberculosis* YopD mutants that genetically separate effector protein translocation from host membrane disruption. Mol Microbiol 96(4):764–778. https://doi.org/10.1111/mmi.12970
85. Aili M, Isaksson EL, Carlsson SE, Wolf-Watz H, Rosqvist R, Francis MS (2008) Regulation of *Yersinia* Yop-effector delivery by translocated YopE. Int J Med Microbiol 298(3-4):183–192
86. Coleman MA, Cappuccio JA, Blanchette CD, Gao T, Arroyo ES, Hinz AK, Bourguet FA, Segelke B, Hoeprich PD, Huser T, Laurence TA, Motin VL, Chromy BA (2016) Expression and association of the *Yersinia pestis* translocon proteins, YopB and YopD, are facilitated by nanolipoprotein particles. PLoS One 11(3): e0150166. https://doi.org/10.1371/journal.pone.0150166
87. Hur J, Kim K, Lee S, Park H, Park Y (2017) Melittin-induced alterations in morphology and deformability of human red blood cells using quantitative phase imaging techniques. Sci Rep 7(1):9306. https://doi.org/10.1038/s41598-017-08675-7
88. Francis MS, Amer AA, Milton DL, Costa TR (2017) Site-directed mutagenesis and its application in studying the interactions of T3S components. Methods Mol Biol 1531:11–31. https://doi.org/10.1007/978-1-4939-6649-3_2
89. Bogdanova A, Kaestner L (2018) The red blood cells on the move! Front Physiol 9:474. https://doi.org/10.3389/fphys.2018.00474
90. Bertani G (2004) Lysogeny at mid-twentieth century: P1, P2, and other experimental systems. J Bacteriol 186(3):595–600
91. Lobo AL, Welch RA (1994) Identification and assay of RTX family of cytolysins. Methods Enzymol 235:667–678. https://doi.org/10.1016/0076-6879(94)35180-5
92. Shaw RK, Daniell S, Ebel F, Frankel G, Knutton S (2001) EspA filament-mediated protein translocation into red blood cells. Cell Microbiol 3(4):213–222
93. Kotlarz A, Tukaj S, Krzewski K, Brycka E, Lipinska B (2013) Human Hsp40 proteins, DNAJA1 and DNAJA2, as potential targets of the immune response triggered by bacterial DnaJ in rheumatoid arthritis. Cell Stress Chaperones 18(5):653–659. https://doi.org/10.1007/s12192-013-0407-1
94. Tang Y, Romano FB, Brena M, Heuck AP (2018) The *Pseudomonas aeruginosa* type III secretion translocator PopB assists the insertion of the PopD translocator into host cell membranes. J Biol Chem 293:8982. https://doi.org/10.1074/jbc.RA118.002766

# Chapter 16

# Analysis of Inflammasome Activation in Response to *Yersinia* Infection by Fluorescence Microscopy Detection of Active Caspase-1 Puncta

**Erin E. Zwack and Igor E. Brodsky**

## Abstract

The type of cell death triggered by a particular environmental stimulus influences the outcome of infection or inflammatory disease processes. The ability to identify the cell death pathway that is activated in response to infection is essential for understanding the pathogenesis and host response to infection. Activation of the cysteine protease caspase-1 in various inflammasome complexes indicates that cells are undergoing pyroptosis, a regulated, proinflammatory cell death. Inflammasome assembly and caspase activation can be measured by various methods ranging from detection of inflammasome-dependent cell death, cytokine secretion, cleavage of caspase-1, or the formation of "puncta" within the cell that contain inflammasome components, such as caspase-1 or the adapter protein ASC. Here we describe a method for detecting caspase-1 activation on a single cell level in the context of infection by the Gram-negative pathogen *Yersinia* using immunofluorescence microscopy. We previously used this approach to quantify caspase-1 puncta formation in cells containing *Yersinia* translocon components (Zwack et al., MBio 6:e02095–14, 2015). This is a modification of methods used previously by Broz et al. (Cell Host Microbe 8:471–483, 2010) and Case and Roy (MBio 2:e00117-11, 2011). By taking a microscopy-based approach that allows us to quantify puncta as well as other cell-biological features of infection (i.e., number of bacteria associated with a particular cell; levels of bacterial effector or translocon proteins in caspase-1 puncta–containing cells; or levels or localization of host cellular proteins), we can better quantify the heterogeneity between cells undergoing pyroptosis and cells that are not under the same infection conditions. These approaches have the potential to generate hypotheses that can enable further mechanistic insight into activation of pyroptosis in response to bacterial infection.

**Key words** Inflammasome, Caspase-1, Type III secretion system, Puncta

## 1 Introduction

As our scientific understanding of cellular biology continues to expand, researchers have found that how a cell dies plays an important role in how our immune system responds. Following the original discovery of apoptosis as a noninflammatory programmed cell death [1], scientists have identified new types of programmed cell death with unique features and outcomes [2] as well as

Viveka Vadyvaloo and Matthew B. Lawrenz (eds.), *Pathogenic Yersinia: Methods and Protocols*, Methods in Molecular Biology, vol. 2010, https://doi.org/10.1007/978-1-4939-9541-7_16, 

discovered that the insult that activates the cell death pathway can change the outcome of the death pathway, such as proinflammatory apoptosis in response to infection [3]. The three main programmed cell death pathways currently identified are apoptosis—involving caspases 3, 7, 8, and 9; pyroptosis—involving caspase-1 and/or caspase-11; and necroptosis—involving RIPK3, RIPK1, and MLKL [1, 4, 5]. As the molecules released by the cells and how the cellular components are broken down differ depending on the type of cell death activated in the cell, the ability to distinguish among the types of cell death is critical to understanding why and how the host responds to different infections and stresses. In this chapter, we focus on identifying one of the defining features of pyroptosis: activation of caspase-1. Pyroptotic cell death occurs when inflammasomes, multiprotein complexes that include caspase-1, form in response to infection or other stress. Caspase-1 is a member of the cysteine–aspartic acid protease family. It exists in an inactive proform until it undergoes proteolytic cleavage that produces a small p10 subunit and a larger p20 subunit. Inside these inflammasomes, caspase-1 undergoes proteolytic cleavage into its active form. This activation of caspase-1 leads to both the processing of IL-1β and IL-18, the secretion of these cytokines and IL-1α, and the death of the cell [4]. Activation of caspase-1 during an infection can be determined by Western blot [6]. Whether caspase-1 is required for cell death can be determined with knock-out mice [6, 7]. These assays tell you at a bulk level if pyroptosis is occurring and are important initial steps in identifying the type of cell death involved in response to infection.

By detecting caspase-1 activation on a single cell level, we are able to gather additional mechanistic insights into how pathogens as well as sterile insults trigger pyroptosis. Through the use of fluorescent microscopy, we are able to identify which cells in an infection are undergoing pyroptosis as well as identify other features of these cells (such as number of bacteria associated with a particular cell; levels of bacterial effector or translocon proteins present in cells that contain caspase-1 puncta; or levels or localization of host cellular proteins). Thus, we can better quantify the differences between cells that lead to the activation of the pyroptotic pathway. These approaches have the potential to generate hypotheses that can enable further mechanistic insight into activation of pyroptosis in response to bacterial infection.

Here we describe a technique using FAM-YVAD-FMK to detect cleave caspase-1, which our lab has successfully used as an alternative to the rabbit anti-caspase-1 p10 antibody from Santa Cruz (unpublished data). This is a modification of methods used previously by Broz et al. [8] and Case and Roy [9].

## 2 Materials

### 2.1 Cell Culture and Infection

1. Luria–Bertani (LB) Agar Plates.
2. Incubators—26 and 37 °C for bacteria and 37 °C, 5% $CO_2$ for cell culture and infection.
3. 2× YT: 16 g/L Tryptone, 10 g/L Yeast Extract, 5 g/L Sodium chloride (NaCl).
4. Inducing Media: 2 × YT, 20 mM Sodium Oxalate, 20 mM Magnesium chloride.
5. Shaking incubator.
6. Biological safety cabinet.
7. Mortar and pestle.
8. 500 mL vacuum filter (Nalgene, 0.2 μm).
9. 70% Ethanol: 95% Ethanol, deionized water ($dH_2O$).
10. 50 mL conical tube.
11. 70 μm Cell Strainer.
12. Hemocytometer.
13. 100 mm Petri dishes.
14. Trypan Blue.
15. Dulbecco's Modified Eagle Medium (DMEM).
16. Complete DMEM: Dulbecco's Modified Eagle Medium, 10% Fetal Bovine Serum (FBS), 10 mM HEPES. Store at 4 °C.
17. L929 cell supernatants. Store at −80 °C (*see* **Note 1**).
18. Macrophage media: Complete DMEM, 30% L929 cell supernatant. Add 100 mL of L929 supernatant to 234 mL of complete DMEM and filter sterilize (final concentration of L929 supernatant—30%).
19. Murine bone marrow-derived macrophages (BMDMs) [10]. BMDMs are generated by growing freshly isolated bone marrow from either wild-type or particular genetic knockout mice, as necessary, for 7–9 days in 10 mL macrophage media in sterile non-TC petri dishes. Feed differentiating bone marrow cells with 5 mL fresh macrophage media on day 4 post-isolation. On days 7–9, the 10 cm dish should be 70–90% confluent and adherent macrophages ready to harvest.
20. Sterile 24-well plate.
21. Round coverslips (*see* **Notes 2** and **3**).
22. FAM-YVAD-FMK (FAM-FLICA in vitro active caspase-1 detection kit).
23. Z-YVAD-FMK: Reconstitute in Dimethyl Sulfoxide (DMSO) to a concentration of 20 mM. Store at −20 °C.

### 2.2 Fixing and Staining Cells

1. 4% Paraformaldehyde (PFA): Dilute 16% Microscopy Grade PFA in Phosphate-Buffered Saline (PBS). Store at 4 °C (*see* **Note 4**).
2. 0.2% Triton X-100: Make a 10–20% Triton X-100 stock. Dilute in PBS to 0.2% Triton X-100.
3. 10% Bovine Serum Albumin (BSA): Dissolve BSA in PBS to make a 10% (w/v) solution. Store at 4 °C for up to 2 weeks.
4. 2% BSA: Dilute 10% BSA stock in PBS. Make fresh.
5. Forceps.
6. Parafilm.
7. Primary antibodies against other proteins (*see* **Note 17**).
8. Fluorophore-conjugated secondary antibodies (*see* **Note 19**).
9. Cell and nuclear stains (*see* **Note 20**).
10. Glass Slides (*see* **Note 5**).
11. Mounting Medium (*see* **Note 6**).

### 2.3 Imaging and Analysis

1. Confocal Microscope (*see* **Note 7**).
2. Image Analysis Software.

## 3 Methods

### 3.1 Cell Culture and Infection

1. Day 0: Streak *Yersinia* on LB Agar Plate and place in incubator set at 28 °C for 2 days.
2. Day 2: Inoculate 3 mL of 2× YT with a single colony of *Yersinia*. Place culture shaking at 28 °C overnight.
3. Place coverslips in a 24-well plate (*see* **Note 8**).
4. Add BMDMs to wells in 1/3 macrophage media in complete DMEM at a concentration of 1.5–2 × $10^5$ cells per well. Place plate in incubator at 37 °C with 5% $CO_2$ (*see* **Note 9**).
5. Day 3: Subculture 50 μL of the overnight *Yersinia* culture in 3 mL of inducing media. Place shaking at 28 °C for 1 h.
6. Switch cultures to shaking at 37 °C for 2 h.
7. Take optical density reading of culture and calculate the amount of culture needed for a Multiplicity of Infection (MOI) of 5.
8. Transfer culture into a microcentrifuge tube and add 1 mL of complete DMEM. Spin for 1 min at max speed in a small benchtop centrifuge. Remove supernatant. Resuspend the pellet in 1 mL of complete DMEM and spin again. Repeat previous wash step one more time.

9. Resuspend in desired volume to infect cells with bacteria at an MOI 5 in a final volume of 40 μL.
10. Replace media on BMDMs with 250 μL of 1/3 macrophage media in complete DMEM.
11. Infect with *Yersinia* for 2 h at an MOI of 5 by adding 40 μL of washed culture and spinning for 5 min at 201 × *g* (*see* **Notes 10** and **11**).
12. One hour before the end of the infection, add 10 μL of 30× FAM-YVAD-FMK (Dilute frozen 150× stock 1/5 in PBS) to appropriate wells. As a no-fluorescence control, add z-YVAD-fmk to a final concentration of 25 μM to a set of cells at the same time as FAM-YVAD-FMK is added to the other wells (*see* **Notes 12** and **13**).

### 3.2 Fixing and Additional Staining of Cells

1. Remove media and wash once with 500 μL PBS before fixing.
2. Layer in 400 μL 4% PFA. Incubate plate in the dark at room temperature for 20 min (*see* **Note 14**). Remove PFA.
3. Wash 2× with 500 μL PBS (*see* **Note 14**).
4. Once the final PBS wash is aspirated off the cells, add 400 μL 0.2% Triton X-100 to the well. Incubate for 10 min in the dark at room temperature to permeabilize the cells. Wash once with 500 μL PBS (*see* **Note 14**).
5. Block cells in the dark for 1 h at room temperature with 500 μL 10% BSA-PBS (*see* **Note 14**).
6. Dilute primary antibodies in 2% BSA-PBS.
7. Remove lid from 24-well plate containing the coverslips. Stretch parafilm across the top of the lid so that the surface is smooth.
8. Place 40 μL of primary antibody dilution in the center of the well outlines. For no primary control wells, place 40 μL of 2% BSA-PBS in the center of the well outlines.
9. Using forceps, grasp the edge of the coverslip and invert it onto the antibody inside the well outline. Place the lid in the tissue culture incubator 37 °C and 5% $CO_2$ for 1 h (*see* **Notes 15–17**).
10. Fill the wells of your original 24-well plate with 500 μL of PBS. Transfer the coverslips into the wells. The side of the coverslip that was exposed to the antibody should face the top of the well. Aspirate the PBS. Wash two more times with PBS (*see* **Note 18**).
11. Make dilution mix of secondary antibodies in 2% BSA-PBS. Replace parafilm on lid with new parafilm. Place 40 μL of secondary antibody dilution in the center of the well outlines. Using forceps, grasp the edge of the coverslip and invert it onto

the antibody inside the well outline. Place the lid in the tissue culture incubator 37 °C and 5% $CO_2$ for 1 h (*see* **Note 19**).

12. Replace PBS in wells with 500 μL of fresh PBS. Transfer the coverslips into the wells. The side of the coverslip that was exposed to the antibody should face the top of the well. Aspirate the PBS. Wash three more times with PBS (*see* **Note 18**).
13. Make stains for nucleus and cytoplasm. Replace parafilm on lid with new parafilm. Place stains in the center of the well outlines. Using forceps, grasp the edge of the coverslip and invert it onto the antibody inside the well outline. Depending on the stain used, perform appropriate incubation (*see* **Note 20**).
14. Replace PBS in wells with 500 μL of fresh PBS. Transfer the coverslips into the wells. The side of the coverslip that was exposed to the antibody should face the top of the well. Aspirate the PBS. Wash three more times with PBS (*see* **Note 18**).
15. Place a small amount of mounting medium on a glass slide. Dip coverslip into $dH_2O$ to rinse off salts from PBS. Gently touch the edge of the coverslip to paper towel to remove $dH_2O$. Invert coverslip cell side down on mounting media. Place at room temperature protected from light overnight to dry mounting media (*see* **Note 21**).

### *3.3 Imaging and Analysis*

1. Set gains using negative control first. Use positive control to adjust gain so that puncta may have slight saturation but are mainly not fully saturated (*see* **Notes 22** and **23**).
2. Image at least five fields per coverslip (*see* **Notes 24–26**).
3. Place slides in a slide box wrapped in tin foil and store at 4 °C (*see* **Note 21**).
4. If images were taken with a confocal microscope and analysis includes localization of puncta or other proteins, analyze images with software that allows quantitation of three-dimensional reconstruction of the *z*-stacks.
5. Single cells should first be identified with a find object function based on the cell stain.
6. Using the positive control, set threshold for fluorescence intensity based on caspase-1 puncta. Apply this threshold to all other images using the find object function to identify caspase-1 puncta (Fig. 1).
7. Export the data for caspase-1 puncta objects contained within cells so that you are provided with information on which cells in the image had puncta. As any additional analysis for other proteins will also provide information on the individual cell, the information can be used to divide analysis into caspase-1 puncta–containing cells and cells without caspase-1 puncta (*see* **Note 27**).

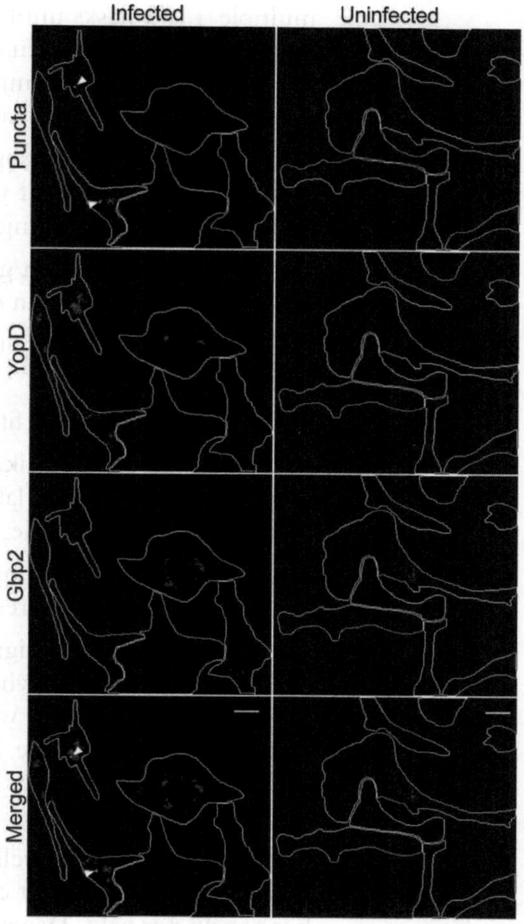


**Fig. 1** Active Caspase-1 puncta staining followed by antibody staining for bacterial and host proteins. Single *z*-stacks of confocal microscopy images of BMDMs infected with $\Delta$*yopEJK* for 2 h or left uninfected. One hour post infection, FAM-FLICA (green) was added to detect active caspase-1. Cells were stained for YopD (blue) and Gbp2 (red). White outlines represent cells as determined by CellMask blue staining. Scale bar represents 10 μm. White arrows indicate puncta

## 4 Notes

1. The following is the protocol for making L929 supernatants. Thaw a fresh batch of L929 cells into 75 $cm^2$ flasks of 100 $cm^2$ plates containing complete RPMI with 10% FBS. When confluent, split into 175 $cm^2$ flasks containing 50 mL complete RPMI 10% FBS. After L929 cells reach confluence, maintain cells in incubator without changing or adding media for 10–12 days. Filter through a 0.2 μm filter flask, combining media from

multiple T175 flasks until the filter flask is full. Filtered L929 supernatant can be frozen and kept at −80 °C indefinitely, but use within 1 year is recommended. Use at 30% to differentiate macrophages from bone marrow.

2. Make sure that the coverslip fits inside the well with some room around the edges so that you can easily remove the coverslips with forceps when staining.

3. Autoclave coverslips in a glass beaker covered with aluminum foil prior to use and open only in a tissue culture hood.

4. To decrease background fluorescence, PFA should be disposed of after 1 month.

5. Make sure that the slide fits the microscope's stage.

6. This protocol is most likely compatible with any mounting medium. However, our lab has only performed it with FluoromountG a water-soluble, nonfluorescing compound.

7. Depending on the information that you are interested in collecting, a wide-field fluorescent microscope can also be used.

8. This protocol was designed using murine bone marrow-derived macrophages, which are naturally adherent cells. If you are using a cell type which is less adherent, you may want to treat coverslips in the 24-well plate with a matrix coating such as poly-L-lysine. This step would need to be performed at least a day in advance.

9. The concentration of cells should allow you to have some separation between most cells while still having a decent number of cells per field. Depending on the size of the cell type that you are using, you may wish to increase or decrease the number of cells per well.

10. If you use a higher MOI, the bacteria replicate to an extent that impairs seeing in the well.

11. Depending on your cell type and what analysis you are interested in besides the presence or absence of caspase-1 puncta, gentamicin should not be added to the well. Phagocytic cells quickly internalize the killed bacteria whose proteins can still be detected by antibodies.

12. All steps once FAM-YVAD-FMK is added should be done while protecting the cells from light.

13. A variation on this protocol has used the discontinued Santa Cruz rabbit anti-casp1 p10 antibody in the primary antibody staining steps and a goat anti-rabbit fluorescent secondary antibody in the secondary antibody staining steps [8, 11, 12]. Theoretically, another caspase-1 antibody should be able to be used for staining if other conditions prevent the use of FAM-YVAD-FMK. In the protocol for detection of caspase-1

puncta using antibodies, BMDM's media should be replaced with 460 μL 1/3 macrophage media in complete DMEM containing 27 μM z-YVAD-FMK immediately after switching *Yersinia* cultures to 37 °C. Infect bacteria at a volume of 40 μL to reach appropriate final volume.

14. The fixing, permeabilizing, and washing steps are done inside the 24-well plate.
15. At this step, you can include other primary antibodies in the mix if their incubation conditions are the same. Otherwise, you can add extra steps as necessary to stain with additional antibodies.
16. The lid is placed in the tissue culture incubator to provide humidity. A different incubator could be used as long as it is humidified.
17. The incubation time and conditions for the primary antibodies were based on antibodies that we have used in conjunction with FAM-YVAD-FMK staining (mouse anti-YopD [13] and rabbit anti-GBP2 [14]). Other incubation times and conditions should also work.
18. Do not allow the coverslips to dry. After washing, cover the coverslips with PBS while preparing the secondary antibody.
19. The incubation time and condition given here for secondary antibodies were determined for goat anti-rabbit Alexa Fluor 568 and goat anti-mouse Alexa Fluor 647 antibodies from Life Technologies (now Fisher). Depending on your secondary antibodies, different conditions may need to be used.
20. This protocol has been successfully performed using HCS CellMask Blue (Fisher) as well as a combination of Hoechst (DNA) and Phalloidin conjugated to Alexa647 (Actin). For CellMask Blue, the lid was placed in a box with wet paper towels at room temperature and incubated for 20–30 min. For the Hoechst/Phalloidin mix, the lid was placed in the tissue culture incubator 37 °C and 5% $CO_2$ for 1 h. The protocol is most likely compatible with other nuclear and cell stains.
21. The company says to read the slides within 24 h. We have determined that slides can be read at least as far as 3 days post staining with FAM-FLICA.
22. Save the microscopy settings that you are using. If you need to take more images from the same coverslip, you will be able to compare to the original images.
23. Puncta are extremely bright and can often be past the point of saturation when gains are set from negative control only.
24. We used a 63× water immersion objective lens.

25. When choosing fields for imaging, use the bright-field setting of the microscope to prevent bias sampling. Select representative fields, but try to make sure that the cells in the field are not touching too much.
26. When using a confocal microscope, make sure that the image includes the bottom and top of all cells in the field.
27. If you are not interested in colocalization of proteins as part of your analysis, a wide-field fluorescence microscope can be employed. Analysis of confocal images can be performed on a "maximum projection," which compresses the individual *z*-planes into a single image. The downstream analysis for determination of caspase-1 puncta should be performed, as previously described.

## References


1. Kerr JF, Wyllie AH, Currie AR (1972) Apoptosis: a basic biological phenomenon with wide-ranging implications in tissue kinetics. Br J Cancer 26:239–257
2. Jorgensen I, Rayamajhi M, Miao EA (2017) Programmed cell death as a defence against infection. Nat Rev Immunol 17:151–164
3. Torchinsky MB, Garaude J, Martin AP, Blander JM (2009) Innate immune recognition of infected apoptotic cells directs T(H)17 cell differentiation. Nature 458:78–82
4. Fink SL, Cookson BT (2005) Apoptosis, pyroptosis, and necrosis: mechanistic description of dead and dying eukaryotic cells. Infect Immun 73:1907–1916
5. Oberst A, Dillon CP, Weinlich R, McCormick LL, Fitzgerald P, Pop C, Hakem R, Salvesen GS, Green DR (2011) Catalytic activity of the caspase-8-FLIP(L) complex inhibits RIPK3-dependent necrosis. Nature 471:363–367
6. Kayagaki N, Warming S, Lamkanfi M, Vande Walle L, Louie S, Dong J, Newton K, Qu Y, Liu J, Heldens S, Zhang J, Lee WP, Roose-Girma M, Dixit VM (2011) Non-canonical inflammasome activation targets caspase-11. Nature 479:117–121
7. Li P, Allen H, Banerjee S, Franklin S, Herzog L, Johnston C, McDowell J, Paskind M, Rodman L, Salfeld J et al (1995) Mice deficient in IL-1 beta-converting enzyme are defective in production of mature IL-1 beta and resistant to endotoxic shock. Cell 80:401–411
8. Broz P, von Moltke J, Jones JW, Vance RE, Monack DM (2010) Differential requirement for Caspase-1 autoproteolysis in pathogen-induced cell death and cytokine processing. Cell Host Microbe 8:471–483
9. Case CL, Roy CR (2011) Asc modulates the function of NLRC4 in response to infection of macrophages by Legionella pneumophila. MBio 2:e00117–11
10. Celada A, Gray PW, Rinderknecht E, Schreiber RD (1984) Evidence for a gamma-interferon receptor that regulates macrophage tumoricidal activity. J Exp Med 160:55–74
11. Zwack EE, Snyder AG, Wynosky-Dolfi MA, Ruthel G, Philip NH, Marketon MM, Francis MS, Bliska JB, Brodsky IE (2015) Inflammasome activation in response to the *Yersinia* type III secretion system requires hyperinjection of translocon proteins YopB and YopD. MBio 6: e02095–14
12. Zwack EE, Feeley EM, Burton AR, Hu B, Yamamoto M, Kanneganti TD, Bliska JB, Coers J, Brodsky IE (2017) Guanylate binding proteins regulate inflammasome activation in response to hyperinjected *Yersinia* translocon components. Infect Immun 85:pii: e00778
13. Noel BL, Lilo S, Capurso D, Hill J, Bliska JB (2009) *Yersinia pestis* can bypass protective antibodies to LcrV and activation with gamma interferon to survive and induce apoptosis in murine macrophages. Clin Vaccine Immunol 16:1457–1466
14. Haldar AK, Saka HA, Piro AS, Dunn JD, Henry SC, Taylor GA, Frickel EM, Valdivia RH, Coers J (2013) IRG and GBP host resistance factors target aberrant, "non-self" vacuoles characterized by the missing of "self" IRGM proteins. PLoS Pathog 9:e1003414

Chapter 17

# Methods for Detection of Pyrin Inflammasome Assembly in Macrophages Infected with *Yersinia* spp.

Natasha P. Medici and James B. Bliska

## Abstract

The *Yersinia* effector proteins YopE and YopT are important bacterial virulence factors that are secreted into infected host cells and can inactivate Rho GTPases, like RhoA, Rac1, and Cdc42. In order to compensate for the consequences of this effect, the host cell can sense RhoA modifications and trigger a proinflammatory reaction to control the infection. This host response, known as pyrin inflammasome assembly, is normally prevented by another important effector, YopM, allowing *Yersinia* to counteract this conserved innate immune response. Once assembled, the pyrin inflammasome can activate caspase-1 via proteolysis, leading to IL-1β secretion and cell death through pyroptosis. Here we describe how to measure pyrin inflammasome assembly, in response to YopE or YopT activities, when macrophages are infected with *yopM* mutant *Yersinia*. Using primary mouse macrophages as host cells, we show how to detect this host response through the downstream events of pyrin dephosphorylation, caspase-1 proteolysis, IL-1β release, and pyroptosis.

**Key words** *Yersinia*, Effectors, Toxins, Inflammasome, Pyrin, Inflammation, Rho GTPases

## 1 Introduction

The genus *Yersinia* consists of 17 species of which three are known to be human pathogens: *Y. pestis*, *Y. pseudotuberculosis*, and *Y. enterocolitica* [1]. Although the pathogenicity levels among those three species are consistently different, they all share a similar trait: the presence of a plasmid-encoded Type III Secretion System (T3SS) as a remarkable virulence factor [2]. This secretion system consists of a syringe-like mechanism responsible for injecting effector proteins into the cytosol of host cells upon close contact. Those effector proteins, known as Yops, have various functions that together will succeed in altering host physiology in order for *Yersinia* to evade the immune system and persist [3].

So far, seven effector Yops are known to be secreted inside cells and among them YopE and YopT are Rho-modifying toxins that utilize different mechanisms to control cellular processes and

Viveka Vadyvaloo and Matthew B. Lawrenz (eds.), *Pathogenic Yersinia: Methods and Protocols*, Methods in Molecular Biology, vol. 2010, https://doi.org/10.1007/978-1-4939-9541-7_17, 

consequently alter homeostasis [3]. YopE mimics Rho GTPase–Activating Proteins (GAPs), keeping RhoA in an inactive state [4]. YopT is a cysteine-protease that cleaves the C-terminus of RhoA, resulting in the release of this protein from the cellular membrane and inactivation [5]. Although the YopE and YopT effectors have different mechanisms to modify Rho GTPases, they both disrupt the actin cytoskeleton, leading to changes in cell morphology, blockage of phagocytosis, and cell mobility [3].

Besides the intentional change in host cell morphology, inactivation of RhoA by YopE or YopT or other bacterial toxins can trigger assembly of the pyrin inflammasome as a compensatory innate immune response [6–8]. Pyrin is a pattern recognition receptor (PRR) that is expressed in macrophages, cytokine-activated monocytes, granulocytes, and serosal and synovial fibroblasts [9]. RhoA-activated kinases PRK1 (PKN1) and PRK2 (PKN2) normally phosphorylate pyrin at serines 205 and 241 in mice and Serines 208 and 242 in humans [10–12]. Phosphorylated pyrin is maintained as an inactive protein by the binding of protein 14-3-3. Upon inactivation of RhoA by bacterial effectors or toxins, phosphorylation of pyrin is lost and protein 14-3-3 is released [10–12]. Consequently, pyrin becomes active and can bind to the adaptor protein ASC, which recruits the cysteine-protease pro--caspase-1, forming the multiprotein complex known as the inflammasome (Fig. 1) [13]. This assembly can mediate auto-proteolysis of pro-caspase-1 into its mature form, which is an important step that leads to the generation and release of active IL-1β and IL-18 cytokines and also cleavage of a cellular protein, gasdermin D (GSDMD) [14]. The generated GSDMD N-terminus fragments relocalizes in the plasma membrane, where it oligomerizes with other N-terminus fragments to form small pores. These pores allow the passage of the cleaved cytokines and also lead to a unique form of cell death, known as pyroptosis [14, 15]. The secretion of proinflammatory cytokines recruits other immune cells to the site of the infection, helping in the clearance of the pathogen. In addition, pyroptotic cell death traps intracellular bacteria and prevents infection of additional cells [16]. Thus, inflammasome function is important for the host to eliminate pathogens [17].

The pyrin inflammasome protects mice against systemic infection by specific pathogens that produce RhoA-inactivating effectors or toxins [7, 8, 18]. The host-adapted pathogen *Yersinia* evolved the effector YopM in order to inhibit YopE- or YopT-triggered pyrin inflammasome formation [7, 8]. YopM counteracts pyrin by hijacking the kinases responsible for keeping this PRR inactivated [7]. Once bound to PRK, YopM increases this kinase's ability to phosphorylate pyrin, consequently negatively regulating the formation of the inflammasome [7]. This leads to immune evasion and persistence of *Yersinia* in the host [7, 8].

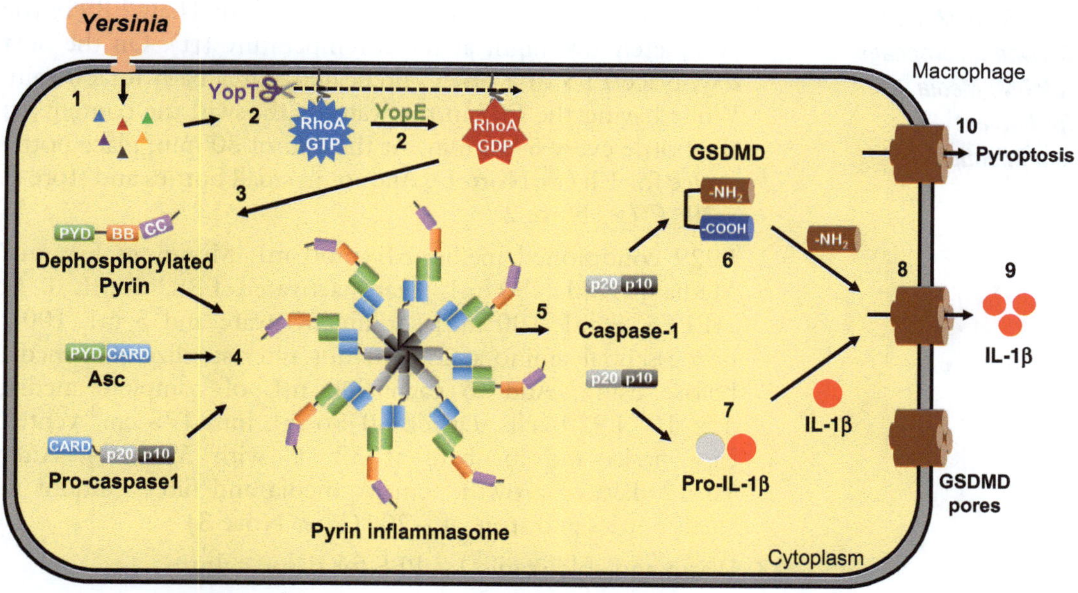


**Fig. 1** *Yersinia* spp. infection triggers the pyrin inflammasome in the absence of YopM. (1) A *Yersinia* bacterium lacking YopM secretes the other six Yop effectors into host cell cytoplasm. (2) YopE and YopT inactivate RhoA by GAP and C-terminal cleavage activities, respectively. (3) The inactivation of RhoA leads to the generation of dephosphorylated pyrin. (4) Dephosphorylated pyrin is active and can recruit Asc and procaspase-1 to form the inflammasome. (5) The inflammasome leads to the generation of mature caspase-1. (6) Caspase-1 can cleave gasdermin-D (GSDMD) to generate GSDMD N-terminal fragments and (7) can also generate mature IL-1β. (8) GSDMD N-terminal fragments can form pores in the membrane, (9) allowing the secretion of IL-1β. (10) GSDMD pores also cause cell death by pyroptosis

To study pyrin inflammasome assembly and function upon host cell infection with *Yersinia*, we optimized different methods that can be used with primary murine macrophages and *yopM* mutant strains. Those methods, used to measure pyrin dephosphorylation, caspase-1 proteolysis, IL-1β release, and pyroptosis, are described below.

## 2 Materials

Prepare all macrophage cell media in a biosafety hood. Macrophage cell media needs to be filter-sterilized before use, while bacterial culture media needs to be autoclaved before use. Keep sterile conditions while performing all experiments and infections to avoid contamination with undesired microorganisms. For media preparation, we suggest the volumes described below. If larger volumes are necessary, adjust accordingly. The material brands described below are suggestions only.

**2.1 Bone Marrow-Derived Macrophage (BMDM) Media for Harvest and Infection**

1. Heat-inactivated Fetal Bovine Serum (FBS): Thaw FBS bottle completely overnight at room temperature (rt). On the next day, place FBS in a 56 °C preheated water bath for 30 min. While having the FBS in the water bath, swirl the contents in the bottle every 5–10 min. At the end of 30 min, place bottle on ice for 1 h (*see* **Note 1**). Aliquot in small bottles and store at −20 °C (*see* **Note 2**).
2. L929 conditioned media: Mix 500 mL Minimum Essential Media (MEM), 50 mL heat-inactivated FBS, 5 mL 1 M HEPES, 5 mL 100 mM sodium pyruvate and 5 mL 100× non-essential amino-acids. Do not filter-sterilize this media before using. Add to each 500 mL of complete media, $4 \times 10^6$ L929 cells. Divide 50–56 mL into 175 $cm^2$ vented caps flasks and incubate at 37 °C with 5% $CO_2$. After 10–12 days of growth, remove media and filter. Aliquot in small bottles and store at −20 °C (*see* **Note 3**).
3. Warm and cold sterile 1× PBS for tissue culture.
4. Nontissue culture-treated petri dishes (100mm).
5. Forceps (autoclaved).
6. Scissors (autoclaved).
7. No. 10 round blade scalpels.
8. 5 mL syringe with 23-gauge (G) needle.
9. Wash Media: In a 50 mL Falcon tube, mix 47.5 mL Dulbecco's Modified Eagle Medium (DMEM) + Glutamax (Gibco®), 2.5 mL heat-inactivated FBS, and 0.5 mL penicillin-streptomycin. Homogenize by inverting the tube 4–6 times (*see* **Note 4**).
10. Macrophage Growth Media 10/30 (MGM 10/30): For 500 mL of media, mix 290 mL DMEM + Glutamax, 5 mL 1 M HEPES (for 10 mM final concentration), 5 mL 100 mM sodium pyruvate (for 1 mM final concentration), 50 mL heat-inactivated FBS (10% total volume of media), and 150 mL of L929 conditioned medium (30% of total volume of media) (*see* **Note 5**) [7].
11. cDMEM media: For 500 mL of media, mix 440 mL of DMEM + Glutamax, 5 mL 1 M HEPES, 5 mL 100 mM sodium pyruvate, and 50 mL heat-inactivated FBS (*see* **Note 5**) [7].
12. MGM 10/10 (infection media): Mix one part of MGM 10/30 to two parts of cDMEM. Make the necessary volume for the experiment (*see* **Note 5**).
13. O26:B6 *Escherichia coli* LPS (Sigma®): Dilute 1 mg LPS lyophilized powder in 10 mL of sterile 1× phosphate-buffered saline (PBS) to obtain a stock solution of 100 μg/mL. Keep 1 mL working stock at 4 °C and store the other 9 mL (divided into 1.5 mL tubes) at −20 °C [19] (*see* **Note 6**).
14. Cold Hank's Balanced Salt Solution (HBSS) for cell culture.

#### 2.2 Bacterial Media, Strains, and Intoxication Conditions

1. Luria broth: For 1 L of media, add 10 g Bacto™ tryptone, 5 g Bacto™ yeast-extract, 10 g sodium chloride (NaCl). Add one sodium hydroxide (NaOH) pellet to adjust media to a pH around 7–8 and then autoclave to sterilize.
2. 0.2 M Sodium Oxide (NaOx): For 1 L, dissolve 26.8 g NaOx into 500 mL double-distilled water. Bring the volume up to 1 L. Divide into 100 mL aliquots and autoclave.
3. 0.2 M Magnesium Chloride ($MgCl_2$): For 1 L, dissolve 40.66 g into 500 mL double-distilled water. Bring the volume up to 1 L. Divide into 100 mL aliquots and autoclave.
4. *Y. pseudotuberculosis yopM* mutant strains [7]: 32777*ΔyopM* ($YopE^+$, $YopT^+$ and $YopM^-$), 32777 *yopE*$^{R144A}$ *ΔyopM* ($YopE^-$, $YopT^+$ and $YopM^-$), 32777 *yopT*$^{C139A}$ *ΔyopM* ($YopE^+$, $YopT^-$ and $YopM^-$). For negative controls: 32777 WT and 32777 *yopE*$^{R144A}$ *yopT*$^{C139A}$ *ΔyopM* ($YopE^-$, $YopT^-$ and $YopM^-$) (*see* **Note** 7).
5. *Clostridium difficile* toxin B (TcdB) (List Biologicals®): Spin down toxin tube for 30–45 s at 20,000 × *g*. Reconstitute 20 μg of TcdB into 100 μL of sterile double-distilled water to obtain a 200 μg/mL stock solution. Store at 2–8 °C (*see* **Note 5**).
6. Phase-contrast light microscope.

#### 2.3 Cell Lysate Analysis

1. Radioimmunoprecipitation assay (RIPA) lysis buffer: For 10 mL of RIPA, mix 500 μL 1 M Tris–HCl pH 8.0 (50 mM final concentration), 1.5 mL 1 M NaCl solution (150 mM final concentration), 100 μL NP-40, 0.05 g deoxycholic acid (DOC) and 0.01 g sodium dodecyl sulfate (SDS). Dissolve all components. Complete volume up to 10 mL. Aliquot and store at −20 °C (*see* **Note 8**).
2. Quantification of protein by Bicinchoninic acid (BCA) assay: Follow BCA Protein Assay Kit protocol recommendations. To create standard reagents, dilute Bovine Serum Albumin (BSA) into RIPA for more accurate results. Follow the microplate procedure for cell lysate analysis. Sample cell lysates can be diluted up to 1:8 in RIPA for the experiment. Do standard BSA curve and samples dilution in duplicate (*see* **Note 9**).
3. 10% SDS-Polyacrylamide gel: To make resolving gel, add 3.3 mL 30% acrylamide mix, 2.5 mL 1.5 M Tris-HCl pH 8.8, 100 μL 10% SDS, 100 μL 10% Ammonium Persulfate (APS), 4 μL Tetraacetylethylenediamine (TEMED). Add 4 mL of double-distilled sterile water to complete a 10 mL final volume. To make stacking gel, add 670 μL 30% acrylamide mix, 500 μL 1.5 M Tris-HCl pH 6.8, 40 μL 10% SDS, 40 μL 10% APS, 4 μL TEMED. Add 2.7 mL of double-distilled sterile water to complete a 4 mL final volume.

4. PVDF membranes (*see* **Note 10**).
5. 4–12% SDS-Polyacrylamide gradient gel.
6. Running buffer: To make a 10× solution, add 30.3 g Tris-HCl Base, 144.2 g Glycine, and 10 g SDS to 300 mL of double-distilled water. After dissolving the components completely, add 700 mL of water to make 1 L. Before performing electrophoresis, do a 1:10 dilution in water to make 1× running buffer.
7. 1× 4-Morpholineethane sulfonic acid (MES) running buffer: To make a 500 mL 1× MES running buffer, mix 25 mL NuPAGE™ 20× MES SDS running buffer with 475 mL double-distilled water. After assembling running apparatus, add 500 μL of NuPAGE™ antioxidant (Thermo Fisher Scientific) into middle chamber.
8. Transfer buffer: To make a 10× solution, add 145 g Glycine and 29 g Tris-HCl base to 300 mL of double-distilled water. After dissolving the components completely, add 700 mL of water to make 1 L. Before performing protein transfer to membrane, make 1× transfer buffer by adding 100 mL 10× transfer buffer, 700 mL double distilled water, and 200 mL methanol (*see* **Note 11**).
9. Plate shaker.
10. Tris-HCl Buffered Saline-Tween (TBS): To prepare 10× solution, add 26.5 g Tris-HCl Base, 44.4 g Tris–HCl, 80 g NaCl, and 2 g potassium chloride (KCl) to 300 mL of double-distilled water. After dissolving all components, add 700 mL of water to complete 1 L. To make 1× TBS-T solution, mix 100 mL of 10× TBS with 900 mL of water (1 L). Add 1 mL of Tween®-20 for a final concentration of 0.1% (*see* **Note 12**).
11. Blocking solution: In a 50 mL conical tube add 4 g powdered milk to 50 mL 1× TBS-T. Mix until the clumps dissolve. Keep it at 4 °C.
12. Amersham ECL detection blotting reagent kit (GE Healthcare®) (*see* **Note 13**).

### *2.4 Primary and Secondary Antibodies*

1. Rabbit monoclonal antimouse total pyrin antibody (Abcam®—ab195975): Divide 100 μL into 10 μL aliquot and store at −20 °C for long term (*see* **Notes 14** and **15**).
2. Rabbit monoclonal antimouse S205 phosphorylated pyrin antibody (Abcam®—ab201784): Divide 100 μL into 10 μL aliquot and store at −20 °C for long term (*see* **Notes 14** and **15**).
3. Rabbit polyclonal antimouse β-actin antibody (Cell signaling®—#4967): Aliquoting is not necessary. Store at −20 °C for long term.

4. Rabbit polyclonal antimouse caspase-1 antibody: Antibody provided by Dr. Gabriel Nuñez. Use as previously described [7]. Store at −80 °C for long term. Keep a working stock at −20 °C for short term (*see* **Notes 14** and **16**).
5. Polyclonal goat antirabbit IgG (H + R) antibody conjugated with horseradish peroxidase (HRP): Rehydrate antibody with indicated volume of water and glycerol (1:1), for a 50% final glycerol volume. Aliquot and store at −20 °C for long-term usage (*see* **Notes 14** and **17**).

### 2.5 Cell Supernatant Analysis

1. Quantikine ELISA Mouse IL-1β/IL-1F2 Immunoassay (R&D Systems®)—Use this kit to measure IL-1β concentrations in the cell culture supernatant. Follow kit instructions carefully (*see* **Note 18**).
2. 96-well microplate for LDH assay.
3. CytoTox 96® Non-Radioactive Cytotoxicity Assay (Promega®)—Use this kit to measure lactate dehydrogenase (LDH) release. Follow kit instructions carefully (*see* **Note 18**).
4. Microplate reader (*see* **Note 19**).
5. Plate shaker.

## 3 Methods

### 3.1 BMDM Isolation and Culture Conditions

1. The isolation of BMDMs should be performed as previously reported [7]. Remove the femurs from 8-week-old WT C57BL/6 mice and clean off muscle tissue as completely as possible using scalpel. Cut ends of the bones and flush them with Wash Media (5 mL per femur) using a 5 mL syringe with a 23 G needle, in a petri dish. Break up clumps of bone marrow gently by pipetting up and down. The solution should become turbid.
2. Transfer the bone marrow cells in Wash Media into a 15 mL conical tube and in a centrifuge, spin for 10 min at 112 × *g* (rt). This is considered the first wash. Once the centrifugation is done, discard the supernatant and resuspend the pellet into 10 mL of Wash Media by pipetting gently up and down. Spin for 5 min at 112 × *g*. This is the second wash. Repeat the last wash one more time, for a total of three washes.
3. After washing, resuspend the pellet with 10 mL of the media MGM 10/30 (5 mL for each femur). Split the resuspension into four 100 mm non-tissue culture dishes (2.5 mL each) and add 7.5 mL of fresh MGM 10/30 to complete a total of 10 mL per dish.

4. Incubate the cells at 37 °C and 5% $CO_2$ for 4 days. On day 4, check cells for viability and contamination. The cells should be adhered to the bottom and have macrophage-like morphology. Remove old media and add 10 mL of fresh MGM 10/30. Return to incubator until day 7.
5. On day 7, after checking for viability, aspirate media and add 5 mL of cold PBS to each dish to remove debris. Shake it slowly and aspirate immediately. Add 10 mL of cold PBS to each dish and incubate them on ice for 10–25 min. Do not incubate the cells on ice for more than 30 min.
6. After incubation, dislodge BMDMs from dish by pipetting up and down. The solution should turn opaque due to the presence of the cells. Once the bottom of the dish is clean, transfer the suspended cells into a 50 mL conical tube and spin (rt) on a centrifuge for 5 min at 112 × *g*.
7. Gently aspirate PBS away from the pellet and add 20 mL of media MGM 10/10 (5 mL per dish) and break up all the clumps. Use a hemocytometer to count the cells and plate the BMDMs as needed. The density on a 6-well plate should be $0.8 \times 10^6$ cells/well in a total volume of 3 mL/well and in a 24-well plate should be $1.5 \times 10^5$ cells/well in a total volume of 1 mL/well. Incubate cells at 37 °C with 5% $CO_2$.
8. One hour after cell plating, add *Escherichia coli* LPS to a final concentration of 100 ng/mL to prime the cells and incubate as in Subheading 3.1, **step** 7 until the next day (day of infection) (*see* **Note 6**).

### 3.2 BMDM Infections

1. Inoculate LB broth with *Y. pseudotuberculosis yopM* mutant strains and controls. Allow them to grow overnight at 28 °C (*see* **Note 7**).
2. The following day, dilute overnight cultures 1:40 in LB broth containing 20 mM sodium oxalate and 20 mM of $MgCl_2$ (final concentration). Grow them for 1 h at 28 °C, and then shift the cultures to 37 °C for 2 h, to allow production of the T3SS and secretion of the Yops.
3. Take 1.5 mL of the cultures, spin down (rt) in a microcentrifuge tube at 20,000 × *g* for 5 min, and aspirate the supernatant. Resuspend the pellet with warm 1.5 mL 1× PBS and take 800 μL to measure the $OD_{600}$ ($OD_{600}$ of $1 = 1.778 \times 10^9$ bacteria/mL). Dilute the bacteria into warm serum-free MGM 10/10 at a Multiplicity of Infection (MOI) of 30 using 15 mL conical tubes for each infection [7].
4. Use a phase-contrast light microscope to check for macrophage LPS-induced activation and sterility. Remove MGM 10/10 containing LPS from macrophages and wash with 1 mL of warm 1× PBS. Remove the PBS wash immediately and add

3 mL of media containing bacteria from **step 3** in appropriate wells. At least two wells should be incubated with serum-free MGM 10/10 only for uninfected control and for maximum LDH release control (described on Subheading 3.6). For a positive control, use 0.2 μg/mL final concentration of purified *C. difficile* TcdB in serum-free MGM 10/10 (*see* **Note 5**).

5. After addition of media with bacteria to LPS-primed BMDMs, spin plates for 5 min at 112 × *g* (rt) to facilitate the contact of *Y. pseudotuberculosis* with BMDMs. Incubate plates for 90 min at 37 °C with 5% $CO_2$ [7].
6. After 90 min, check BMDMs by phase-contrast light microscopy to confirm that the bacterial infection has resulted in YopE- and YopT-mediated cell rounding and collect 1.5 mL supernatant from each well, with the exception of the LDH control. Spin supernatant in microcentrifuge tube for 10 min at 20,000 × *g* (4 °C) and transfer 1 mL to a new tube. Store at −20 °C.
7. Remove the remaining culture media from plates and wash the BMDMs once with 1 mL of cold HBSS. Add 100 μL of RIPA lysis buffer with protease inhibitor cocktail to coat each well and incubate plates on ice for 10–15 min (*see* **Note 20**).
8. Remove the plates from ice and use a pipet tip to scrape the bottom of plates to suspend remaining cell debris. Collect the buffer with lysed cells and spin in a microcentrifuge tube for 20 min at 20,000 × *g* (4 °C). Transfer cell lysate supernatant to a new tube and store it at −20 °C.
9. The LDH control well should be kept at −80 °C until used (*see* **Note 21**).

### 3.3 Western Blot for Detection of Total and Phosphorylated Pyrin

1. Calculate total protein concentration in cell lysate by using a protein assay standard curve (described in Subheading 2.3, **item 2**).
2. Resolve the samples on a 10% SDS-PAGE gel at 200 V. Load between 3 and 15 μg for best resolution and visualization of proteins.
3. After electrophoresis, transfer proteins to a PVDF membrane for 1 h at 350 mA.
4. Block membrane for 1 h in Blocking solution.
5. Probe membrane overnight by shaking it at 4 °C with rabbit monoclonal antibody against mouse pyrin in Blocking solution in a 1:1000 dilution for detection of total pyrin (*see* **Notes 15** and **20**).
6. For detection of phosphorylated (inactive) pyrin, probe membrane overnight by shaking it at 4 °C with rabbit monoclonal

antibody against mouse S205 phosphorylated pyrin, in Blocking solution in a 1:1000 dilution (*see* **Notes 15** and **20**).

7. Wash excess primary antibody from membrane with 1× TBS-T four times and probe with secondary goat antirabbit antibody conjugated to HRP, in a 1:10,000 dilution. This secondary antibody can be used for detection of both S205 phosphorylated pyrin and total pyrin.
8. After 1 h, remove secondary antibody and wash membrane with 1× TBS-T four times. Visualize desired proteins by using ECL chemiluminescent detection reagent. Signal visualization can be obtained by exposure to X-ray film or through use of a Western Blot Imager. Bands for both phosphorylated (inactive) and total pyrin should be visualized between 100 and 110 kb.
9. As a loading control, a primary rabbit monoclonal antibody against mouse β-actin can be used on the same membranes. Use an antirabbit secondary antibody conjugated to an HRP for visualization, as described in **step 6**.

### 3.4 Western Blot for Detection of Caspase-1 Proteolysis

1. Using the total protein concentration obtained in Subheading 3.1, **step 1**, load between 3 and 15 μg of protein into a 4–12% gradient gel. Resolve the samples by running the gel in 1× MES running buffer at 175 V and transfer to a PVDF membrane as described above.
2. Block membrane for 1 h in Blocking solution and probe overnight for caspase-1 by shaking at 4 °C using anti caspase-1 rabbit polyclonal antibody [7] in Blocking solution in a 1:10,000 dilution.
3. Wash excess primary antibody from membrane with 1× TBS-T three times and probe with secondary polyclonal goat antirabbit IgG (H + R) + HRP in a 1:20,000 dilution. Wash excess secondary antibody with 1× TBS-T four times. Signal visualization can be obtained as in Subheading 3.3, **step 6**. Using the gradient gel and the antibody described in **step 2** enables the visualization of both procaspase-1 and cleaved caspase-1 (*see* **Note 16**).
4. In this case, β-actin can also be used as a loading control. Follow the same steps as summarized on Subheading 3.3, **step** 7.

### 3.5 Supernatant Analysis for IL-1β Release: ELISA Kit

1. Thaw frozen supernatants at room temperature. Spin samples at 20,000 × *g* in a microcentrifuge (rt). Prepare reagents according to the commercial kit protocol. Prepare the Mouse IL-1β standard according to vial label. This generates an 800 pg/mL stock solution, which can be used to produce

twofold dilutions series from 800 to 0 pg/mL. Those dilutions will be part of the standard curve.

2. The kit provides 96-well microplates with strips of wells. Remove excess wells and save for future experiments.
3. Follow the kit protocol. A microplate reader that can be set between 450 and 570 nm is necessary for results collection. Analyze through a software that generates graphs and compares different data sets (Excel, GraphPad Prism, etc.). The standard curve should generate an equation based on absorbance × concentration: this formula should be used to calculate sample concentration (*see* **Note 22**). Graph needs to compare Il-1β release in the different samples.
4. For biological relevance, collect data from three independent experiments, and analyze and compare them through a statistical test (one-way ANOVA, two-way ANOVA, Mann-Whitney, etc.), according to the experiment goals.

### 3.6 Supernatant Analysis for Cell Death Through Pyroptosis: LDH Release Assay Kit

1. Thaw frozen supernatants at room temperature as described in Subheading 3.5, **step 1**.
2. Thaw LDH control well, collect media, and spin in a microcentrifuge tube for 20 min at 20,000 × *g* (4 °C). Have a no-cell control, containing only fresh MGM 10/10 as a negative control.
3. Set up a 96-well flat bottom plate and add 50 μL aliquots from controls and samples. Follow kit protocol. In this case, a microplate reader that analyzes 490 or 492 nm is necessary.
4. Calculate the results according to kit protocol and analyze data through graph-generating software, as explained in Subheading 3.5, **step 4**. In this case, the graph is comparing LDH release percentage in the different samples.
5. For biological relevance, collect data from three independent experiments, and analyze and compare them through a statistical test (one-way ANOVA, two-way ANOVA, Mann-Whitney, etc.), according to the experiment goals.

## 4 Notes

1. Water level in water bath should be sufficient to cover at least half of the bottle of FBS. Ice level should be enough to be above serum level during 1 h.
2. Constant thaw–freeze cycles are not recommended for FBS. Make sure to aliquot necessary amounts for one experiment. Test new batches of FBS before using to make sure it supports BMDM growth and differentiation.

3. Before starting L929 cell culturing, check for cracks or broken 175 $cm^2$ flasks. Also, make sure to have enough flasks for the whole experiment (10–20 flasks). By 10–12 days, L929 cells should be attached to bottom of 175 $cm^2$ flasks. In case the cells start to detach, remove media immediately, transfer to 50 mL conical tubes, and spin down in a centrifuge at 112 × *g* (rt) before filtering, to avoid clogging. Thaw–freeze cycles are not recommended for L929 conditioned media. Make sure to aliquot necessary amounts for one experiment.

4. Wash media mixture should be at room temperature for best bone marrow harvest. Sterilization is not required for this step.

5. FBS is essential for BMDM growth and differentiation. However, during intoxication of cells with TcdB, serum-free media is required. FBS and TcdB compete for the same cell receptors and toxin effects will not be observed [10]. Thus, we recommend that for infections using TcdB as a positive control, **step 4** from Subheading 3.2 is followed. To generate serum-free media, follow same steps for media preparation (*see* Subheading 2.1, **items 4–6**); however, *do not add FBS*. Adjust DMEM volume accordingly. Do not change other components concentrations. Absence of FBS on *Yersinia* spp. infections should not affect results. Serum-free media and FBS-containing media should be warmed to 37 °C before application.

6. The LPS stock generated (100 μg/mL) will allow a 1:1000 dilution for a final concentration of 100 ng/mL. Due to the small volume necessary to prime the cells in each well (3 μL of LPS if using 3 mL/well), we recommend a 1:10 dilution of the LPS in PBS as a first step. From this dilution, use a volume equivalent to 1:100 dilution into the wells, to have a 100 ng/mL final concentration of LPS.

7. Pyrin inflammasome is triggered in the absence of YopM, but in the presence of YopE and YopT. It is important to use mutants lacking *yopM* in order to observe the inflammasome effects. Although the described protocol was established in *Y. pseudotuberculosis*, other pathogenic *Yersinia* species that can produce the T3SS, such as *Y. pestis* [19] and *Y. enterocolitica*, are suitable for those experiments.

8. Other types of cell lysis buffers, such as NP-40 lysis buffer [7], can be used for those experiments. The cell lysis buffer should be carefully chosen depending on the protein analyzed and its cellular location. RIPA thaw–freeze cycles are not recommended. Make sure to create aliquots that have enough lysis buffer for one experiment.

9. BSA for standard curve and samples should be diluted in the cell lysis buffer used to generate cell lysate. Dilutions of 1:8 are recommended for analysis of cell lysate samples. In case 1:8

sample dilutions have absorbance values lower than absorbance value of lowest standard curve dilution, repeat experiment by making less diluted samples (1:6, 1:4, 1:2, etc.). Although BCA protein assay is explained in this protocol, other protein quantification assays, such as Bradford, can be used.

10. PVDF membranes need to be activated prior to protein transfer. To do that, place membrane into Methanol. The volume should be enough to cover the membrane. Leave it for at least 10 min. Wash three times with double-distilled water and assemble transfer apparatus.
11. While preparing 1× Transfer buffer, add the methanol last. This will prevent precipitation of buffer components.
12. 1× TBS-T should have Tween concentrations between 0.05% and 0.1% for best results.
13. The ECL reagent kit comes with two reagents, A and B. Mix them in a 1:1 dilution immediately before using film or Western Blot Imager. The reagent is light-sensitive and might lose effectiveness if mixed in advance. For membranes of size 7 cm × 10 cm or smaller, no more than 500 μL ECL mix reagent is necessary to visualize proteins. If the membrane is bigger than the sizes specified, increase volume of reagent.
14. Upon arrival of antibodies, spin down tubes in microcentrifuge tube at 20,000 × *g* (rt) before rehydration and/or aliquoting. Avoid thaw–freeze cycles.
15. Antibody against total pyrin binds pyrin specifically between amino acids 50–450. This allows detection of both active and inactive pyrins. Antibody against S205 phosphorylated pyrin binds only pyrin that is specifically phosphorylated at serine 205. This allows detection of inactive pyrin only.
16. Antibody against caspase-1 can bind to precursor procaspase-1 (50 kDa) and cleaved caspase-1 at p20 subunit (20 kDa).
17. Rehydration of secondary antibody can be made in water only as well. For extended storage in this case, store at −80 °C.
18. Note that thawing and freezing cycles of cell supernatants should be avoided. It is recommended that both ELISA and LDH assays are performed in the same day.
19. Microplate reader needs to be capable of recording absorbances 490 or 492 nm (LDH assay) and 450 and 540 or 570 nm (ELISA).
20. If using the antibodies described in Subheading 2.4, **items 1** and **2**, avoid the addition of phosphatase inhibitors to the cell lysis buffer. The phosphatase inhibitor treatment can change the binding properties of the total pyrin antibody, leading to suboptimal results. If phosphatase inhibitor treatment is

required, use a rabbit monoclonal antibody against mouse total pyrin that recognizes amino acids 500–750 (Abcam® ab 214772).

21. The freezing and thawing of macrophages generates maximum cell death and can be used as a positive control during the LDH release analysis.
22. For accurate results, make sure that the standard curve has $R^2$ equivalent or higher than 0.95.

## Acknowledgments

The authors thank Nicole Loeven and Amrapali Ghosh for editing the manuscript. The preparation of this publication was supported by the NIH under award R01AI099222 (JB) and by Science Without Borders/CAPES—Brazil (NM).

### References

1. Cornelis GR, Wolf-Watz H (1997) The *Yersinia* Yop virulon: a bacterial system for subverting eukaryotic cells. Mol Microbiol 23 (5):861–867. https://doi.org/10.1046/j.1365-2958.1997.2731623.x
2. Cornelis GR (2002) *Yersinia* type III secretion: send in the effectors. J Cell Biol 158 (3):401–408. https://doi.org/10.1083/jcb.200205077
3. Viboud GI, Bliska JB (2005) *Yersinia* outer proteins: role in modulation of host cell signaling responses and pathogenesis. Annu Rev Microbiol 59:69–89. https://doi.org/10.1146/annurev.micro.59.030804.121320
4. Black DS, Bliska JB (2000) The RhoGAP activity of the *Yersinia pseudotuberculosis* cytotoxin YopE is required for antiphagocytic function and virulence. Mol Microbiol 37(3):515–527
5. Shao F, Vacratsis PO, Bao Z, Bowers KE, Fierke CA, Dixon JE (2003) Biochemical characterization of the *Yersinia* YopT protease: cleavage site and recognition elements in Rho GTPases. Proc Natl Acad Sci U S A 100 (3):904–909. https://doi.org/10.1073/pnas.252770599
6. Xu H, Yang J, Gao W, Li L, Li P, Zhang L, Gong Y-N, Peng X, Xi JJ, Chen S, Wang F, Shao F (2014) Innate immune sensing of bacterial modifications of Rho GTPases by the Pyrin inflammasome. Nature 513 (7517):237–241. https://doi.org/10.1038/nature13449
7. Chung LK, Park YH, Zheng Y, Brodsky IE, Hearing P, Kastner DL, Chae JJ, Bliska JB (2016) The *Yersinia* virulence factor YopM hijacks host kinases to inhibit type III effector-triggered activation of the pyrin inflammasome. Cell Host Microbe 20 (3):296–306. https://doi.org/10.1016/j.chom.2016.07.018
8. Ratner D, Orning MPA, Proulx MK, Wang D, Gavrilin MA, Wewers MD, Alnemri ES, Johnson PF, Lee B, Mecsas J, Kayagaki N, Goguen JD, Lien E (2016) The *Yersinia pestis* effector YopM inhibits pyrin inflammasome activation. PLoS Pathog 12(12):e1006035. https://doi.org/10.1371/journal.ppat.1006035
9. Chae JJ, Wood G, Masters SL, Richard K, Park G, Smith BJ, Kastner DL (2006) The B30.2 domain of pyrin, the familial Mediterranean fever protein, interacts directly with caspase-1 to modulate IL-1beta production. Proc Natl Acad Sci U S A 103 (26):9982–9987. https://doi.org/10.1073/pnas.0602081103
10. Gao W, Yang J, Liu W, Wang Y, Shao F (2016) Site-specific phosphorylation and microtubule dynamics control Pyrin inflammasome activation. Proc Natl Acad Sci U S A 113(33): E4857–E4866. https://doi.org/10.1073/pnas.1601700113
11. Masters SL, Lagou V, Jéru I, Baker PJ, Van Eyck L, Parry DA, Lawless D, De Nardo D, Garcia-Perez JE, Dagley LF, Holley CL, Dooley J, Moghaddas F, Pasciuto E, Jeandel P-Y, Sciot R, Lyras D, Webb AI, Nicholson SE, De Somer L, van Nieuwenhove E, Ruuth-Praz J, Copin B, Cochet E, Medlej-Hashim M,

Megarbane A, Schroder K, Savic S, Goris A, Amselem S, Wouters C, Liston A (2016) Familial autoinflammation with neutrophilic dermatosis reveals a regulatory mechanism of pyrin activation. Sci Transl Med 8(332):332ra345

12. Park YH, Wood G, Kastner DL, Chae JJ (2016) Pyrin inflammasome activation and RhoA signaling in the autoinflammatory diseases FMF and HIDS. Nat Immunol 17 (8):914–921. https://doi.org/10.1038/ni.3457. http://www.nature.com/ni/journal/v17/n8/abs/ni.3457.html#supplementary-information
13. Shin S, Brodsky IE (2015) The inflammasome: learning from bacterial evasion strategies. Semin Immunol 27(2):102–110. https://doi.org/10.1016/j.smim.2015.03.006
14. Shi J, Gao W, Shao F (2017) Pyroptosis: gasdermin-mediated programmed necrotic cell death. Trends Biochem Sci 42 (4):245–254. https://doi.org/10.1016/j.tibs.2016.10.004
15. Kovacs SB, Miao EA (2017) Gasdermins: effectors of pyroptosis. Trends Cell Biol 27 (9):673–684. https://doi.org/10.1016/j.tcb.2017.05.005
16. Jorgensen I, Miao EA (2015) Pyroptotic cell death defends against intracellular pathogens. Immunol Rev 265(1):130–142. https://doi.org/10.1111/imr.12287
17. von Moltke J, Ayres JS, Kofoed EM, Chavarría-Smith J, Vance RE (2013) Recognition of bacteria by inflammasomes. Annu Rev Immunol 31(1):73–106. https://doi.org/10.1146/annurev-immunol-032712-095944
18. Aubert Daniel F, Xu H, Yang J, Shi X, Gao W, Li L, Bisaro F, Chen S, Valvano Miguel A, Shao F (2016) A burkholderia type VI effector deamidates rho GTPases to activate the pyrin inflammasome and trigger inflammation. Cell Host Microbe 19(5):664–674. https://doi.org/10.1016/j.chom.2016.04.004
19. Schoberle TJ, Chung LK, McPhee JB, Bogin B, Bliska JB (2016) Uncovering an important role for YopJ in the inhibition of caspase-1 in activated macrophages and promoting *Yersinia pseudotuberculosis* virulence. Infect Immun 84(4):1062–1072. https://doi.org/10.1128/iai.00843-15

# Index




Viveka Vadyvaloo and Matthew B. Lawrenz (eds.), *Pathogenic Yersinia: Methods and Protocols*, Methods in Molecular Biology, vol. 2010, https://doi.org/10.1007/978-1-4939-9541-7, © Springer Science+Business Media, LLC, part of Springer Nature 2019

## T



## X



## Y

MIX
Papier aus verantwortungsvollen Quellen
Paper from responsible sources
FSC® C105338

If you have any concerns about our products,
you can contact us on
**ProductSafety@springernature.com**

In case Publisher is established outside the EU,
the EU authorized representative is:
**Springer Nature Customer Service Center GmbH**
**Europaplatz 3, 69115 Heidelberg, Germany**

Printed by Libri Plureos GmbH
in Hamburg, Germany